Springer Complexity

Springer Complexity is an interdisciplinary program publishing the best research and academic-level teaching on both fundamental and applied aspects of complex systems—cutting across all traditional disciplines of the natural and life sciences, engineering, economics, medicine, neuroscience, social and computer science.

Complex Systems are systems that comprise many interacting parts with the ability to generate a new quality of macroscopic collective behavior the manifestations of which are the spontaneous formation of distinctive temporal, spatial or functional structures. Models of such systems can be successfully mapped onto quite diverse "real-life" situations like the climate, the coherent emission of light from lasers, chemical reaction-diffusion systems, biological cellular networks, the dynamics of stock markets and of the Internet, earthquake statistics and prediction, freeway traffic, the human brain, or the formation of opinions in social systems, to name just some of the popular applications.

Although their scope and methodologies overlap somewhat, one can distinguish the following main concepts and tools: self-organization, nonlinear dynamics, synergetics, turbulence, dynamical systems, catastrophes, instabilities, stochastic processes, chaos, graphs and networks, cellular automata, adaptive systems, genetic algorithms and computational intelligence.

The three major book publication platforms of the Springer Complexity program are the monograph series "Understanding Complex Systems" focusing on the various applications of complexity, the "Springer Series in Synergetics", which is devoted to the quantitative theoretical and methodological foundations, and the "Springer Briefs in Complexity" which are concise and topical working reports, case studies, surveys, essays and lecture notes of relevance to the field. In addition to the books in these two core series, the program also incorporates individual titles ranging from textbooks to major reference works.

Series Editors

Understanding Complex Systems

Founding Editor: S. Kelso

Future scientific and technological developments in many fields will necessarily depend upon coming to grips with complex systems. Such systems are complex in both their composition–typically many different kinds of components interacting simultaneously and nonlinearly with each other and their environments on multiple levels–and in the rich diversity of behavior of which they are capable.

The Springer Series in Understanding Complex Systems series (UCS) promotes new strategies and paradigms for understanding and realizing applications of complex systems research in a wide variety of fields and endeavors. UCS is explicitly transdisciplinary. It has three main goals: First, to elaborate the concepts, methods and tools of complex systems at all levels of description and in all scientific fields, especially newly emerging areas within the life, social, behavioral, economic, neuro- and cognitive sciences (and derivatives thereof); second, to encourage novel applications of these ideas in various fields of engineering and computation such as robotics, nano-technology, and informatics; third, to provide a single forum within which commonalities and differences in the workings of complex systems may be discerned, hence leading to deeper insight and understanding.

UCS will publish monographs, lecture notes, and selected edited contributions aimed at communicating new findings to a large multidisciplinary audience.

More information about this series at http://www.springer.com/series/5394

Ted Carmichael · Andrew J. Collins ·
Mirsad Hadžikadić
Editors

Complex Adaptive Systems

Views from the Physical, Natural, and Social Sciences

Springer

Editors
Ted Carmichael
Complex Systems Institute, Department of Software and Information Systems
College of Computing and Informatics,
University of North Carolina at Charlotte
Charlotte, NC, USA

Andrew J. Collins
Department of Engineering Management and Systems Engineering, College of Engineering and Technology,
Old Dominion University
Norfolk, VA, USA

Mirsad Hadžikadić
Complex Systems Institute, Department of Software and Information Systems
College of Computing and Informatics,
University of North Carolina at Charlotte
Charlotte, NC, USA

ISSN 1860-0832 ISSN 1860-0840 (electronic)
Understanding Complex Systems
ISBN 978-3-030-20307-8 ISBN 978-3-030-20309-2 (eBook)
https://doi.org/10.1007/978-3-030-20309-2

This Springer imprint is published by the registered company Springer Nature Switzerland AG
The registered company address is: Gewerbestrasse 11, 6330 Cham, Switzerland

Preface

Most interesting phenomena in natural and social systems include constant transitions and oscillations among their various phases. Biological and human systems rarely stay in a stable, predictable state for long. Randomness, power laws, and human behavior ensure that the future is both unknown and challenging. How do events unfold? When do they take hold? Why do some initial events cause an avalanche while others do not? What characterizes these events? What are the thresholds that differentiate a sea change from a non-event?

Complex Adaptive Systems (CAS) have proven to be a powerful tool for exploring these and other related phenomena. We characterize a general CAS model as having a large number of self-similar agents that: (1) utilize one or more levels of feedback; (2) exhibit emergent properties and self-organization; and (3) produce nonlinear dynamic behavior. Advances in modeling and computing technology have led not only to a deeper understanding of complex systems in many areas, but they have also raised the possibility that similar fundamental principles may be at work across these systems, even though the underlying principles may manifest themselves differently.

This book emerged out of international conferences organized through the Association for the Advancement of Artificial Intelligence (AAAI) Fall Symposia series and the Swarmfest 2017 conference. Our goal was to bring together researchers from diverse fields who study these complex systems using the tools and techniques of CAS and agent-based modeling. In the past, knowledge gained in each domain has remained mostly exclusive to that domain, especially when the disciplines are far apart. It is our belief that by bringing together scholars who study these phenomena, we can leverage a deep knowledge of one domain to gain insight into others.

Charlotte, NC, USA Ted Carmichael
Norfolk, VA, USA Andrew J. Collins
Charlotte, NC, USA Mirsad Hadžikadić

Contents

The Fundamentals of Complex Adaptive Systems

Ted Carmichael and Mirsad Hadžikadić

Abstract Complex Adaptive Systems (CAS) is a framework for studying, explaining, and understanding systems of agents that collectively combine to form emergent, global level properties. These agents can be nearly anything, from ants or bees, to brain cells, to water particles in a weather pattern, to groups of cars or people in a city or town. These agents produce emergent patterns via correlated feedbacks throughout the system, feedbacks that create and fortify a *basin of attraction*: a persistent pattern of behavior that itself is outside of equilibrium. There is also an ever-growing understanding that similar features in complex systems across a diversity of domains may indicate similar fundamental principles at work, and as such there is often utility in using the key features of one system to gain insight into the workings of seemingly distinct fields. Here we also include a brief review of multiple models that attempt to do exactly this, including some of our previous work. Though there is not complete agreement on all aspects and definitions in this field, this introduction also summarizes our understanding of what defines a CAS, including the concepts of complexity, agents, adaptation, feedbacks, emergence, and self-organization; and places this definition and its key features in a historical context. Finally we briefly discuss two of the common biases often found that the tools of CAS can help counteract: the hierarchical bias, assuming a strong top-down organization; and the complexity bias, the tendency to assign complicated features to agents that turn out to be quite simple.

T. Carmichael (✉) · M. Hadžikadić
Department of Software and Information Systems, University of North Carolina at Charlotte, 9201 University City Blvd, Charlotte, NC 28223, USA
e-mail: tedsaid@gmail.com

M. Hadžikadić
e-mail: mirsad@uncc.edu

T. Carmichael
TutorGen, Inc., 1037 S Ft Thomas Ave, Fort Thomas, KY 41075, USA

T. Carmichael et al. (eds.), *Complex Adaptive Systems*, Understanding Complex Systems, https://doi.org/10.1007/978-3-030-20309-2_1

1 Overview

Most interesting collective phenomena in natural and social systems can be described as having stable and persistent states, often outside of equilibrium. The term *basin of attraction* has been used to describe such systems, capturing the idea of correlated feedbacks among the agents of a system that create these identifiable and distinct patterns. These systems are so defined because they are resilient in the face of external forces, but can nevertheless also exhibit tipping points: situations where the stable system finally crosses some threshold, and begins a rapid transition to a new state. These thresholds can be characterized as a qualitative change in system characteristics: a change in sign or abrupt change in magnitude (either enduring or a spike) in the first or second derivative of a system variable.

Threshold effects are found all around us. In economics, this could be movement from a bull market to a bear market; in sociology, it could be the spread of political dissent, culminating in rebellion; in biology, the immune system response to infection or disease as the body moves from sickness to health. Companies, societies, markets, or even humans represent such persistent states that can change rapidly at any time. Both endogenous and exogenous feedbacks can cause sudden, non-linear shifts in system behavior, ensuring that the future of these systems are often unknown and challenging. How do events unfold? When do they take hold? Why do some initial events cause an avalanche of change while others do not? What characterizes system stability and resilience? What are the thresholds that differentiate a sea change from negligible variations?

Complex Adaptive Systems (CAS) has proven to be a powerful framework for exploring thresholds and resilience, and other related phenomena. As the name implies, a CAS is a system of agents that interact among themselves and/or their environment, such that even relatively simple agents with simple rules of behavior can produce complex, emergent behavior. The key to CAS is that the system-level properties generally cannot be understood, or often even defined, at the level of the individual agent description. Therefore, these systems must be studied holistically, as the sum of the agents and their interactions.

1.1 Defining CAS

We characterize a general CAS model as having a significant number of self-similar agents that:

- Utilize one or more levels of feedback;
- Exhibit emergent properties and self-organization;
- Produce non-linear dynamic behavior.

The CAS framework can be used to describe systems that encompass phenomena across many diverse environments and a wide range of disciplines. These systems are present at all scales of inquiry: from the movement of markets and economies to individual knowledge acquisition; from large-scale social interaction to small-scale cellular behavior. Advances in modeling and computing technology have not only led to a deeper understanding of complex systems in many areas but have also raised the possibility that similar fundamental principles may be at work across a wide variety of domains. This idea has led to several multidisciplinary conferences forming to allow the sharing of ideas across domains, including the annual Swarmfest meeting, and The Association for the Advancement of Artificial Intelligence (AAAI) CAS Fall Symposia series, from where the papers in this volume are drawn.

The overriding goal for these conferences is to create synergy and build connections among experts from different domains. Often, complex systems from two distinct fields may seem different on the surface, but have quite similar underlying dynamics. We hypothesize that by modeling complex systems from many different areas, we can start to find the principles that show common causes and common effects across domains. In this way, the known causes and mechanisms in one domain are used to gain insight into the controlling properties of similar effects in other domains. As Neil Johnson writes:

> In particular, the connections between such systems have not been properly explored—particularly between systems taken from different disciplines such as biology and sociology. Indeed it is fascinating to see if any insight gained from having partially understood one system, say from biology, can help us in a completely different discipline, say economics [14, p. 16].

Put another way, Epstein writes:

> Generality, while a commendable impulse, is not of paramount concern to agent-based modelers at this point [10, p. 1602].

And so we believe that by bringing these researchers together, who study different fields but use the same tools and techniques of CAS and Agent-based Modeling (ABM), we can overcome the natural tendency of scholars to work only within their own silos, and encourage fruitful and cross-disciplinary collaborations that successfully draw generalities across domains.

1.2 Common Models Across Diverse Domains

As an illustrative example consider the model found in Midgley, Marks, and Kunchamwar [17], one of numerous examples of using ABM to implement a CAS framework to further understanding of the dynamics found within a particular system. In this work, the authors construct a model that aims to reproduce a typical market structure by utilizing the properties of a supermarket setting. Their model incorporates three types of agents: consumers, retailers, and manufacturers. They have chosen ABM over more traditional methods of model construction that use game theory or analytical equations of system dynamics, due to the power and flexibility of CAS:

> [O]ne can more easily incorporate the existing knowledge about the nature of human-decision-making processes into AB models than into analytical equations. [...] AB models allow a flexibility of representation that is not present in more traditional approaches.

But Midgley et al.'s model was not designed with general applicability in mind. It may be that some of the agent attributes could reasonably be applied to other domains. For example, *chance of observing a store promotion* might be a stand-in for *vision*; *number of best promotions remembered* may be generalized as *memory*; and perhaps *satisfaction threshold* for an agent could represent any state-change threshold for any agent. But it has not been explicitly explored how these translations may be realized, or to what advantage, in a different system. Other attributes, such as *range of advertising levels* or *quarterly increment/decrement to mark-up*, may not have any obvious analogues. Further, the rules governing calculations that utilize these attributes also suffer a lack of generalizability or an explicit method for applying these rules to a new domain.

In [22] the authors present a more general economic model, utilizing only two agents: buyers and sellers. While this work is intended to demonstrate the utility of ABM in this context, it is quite clear that these agents may be easily applicable to many types of markets. However, as with [17], there is no discussion or representation of this models applicability to systems that are outside of economics.

Examples from other domains also follow this common pattern. Vries and Biesmeijer have created an ABM of honeybee foraging [7], which they expanded upon in [8]. While this work is intended to utilize enough flexibility to represent a broad range of variable values found in real-world honeybee colonies, it does not purport to show general adaptability to other fields. Similarly, ABM has been used to develop sophisticated tools for the study of traffic flow under a wide spectrum of environmental factors, such as weather, infrastructure, and changing demographics. Reference [11] describes one such system; but again, limited to only a single domain.

There have also been examples of ideas or concepts of CAS taken from one domain and applied to one or more others. Schelling's classic model on segregation [20] is an example of a fundamental property, one that may be readily applied to many systems, informing models found in sociology, biology, or economics. Flocking behavior has been studied in birds, fish, and crowds of people, and simple analogies between these diverse systems can be drawn [21]. Also, the collective intelligence of ants for determining the shortest path has proven to be useful in the engineering of decentralized flow control, such as in computer networks. In general, these examples illustrate how one system can inform study of another: either by drawing comparisons from one model to another, or by using certain properties found in one model to inform the construction of a second model.

In furthering this idea our prior work has explored using a single CAS tool to replicate key properties of complex systems as found in multiple domains: a single model with multiple applications. This model was developed and used to simulate the growth of cancer and the immune system response; and then used to show similarities in the growth of a social contagion effect in a polity, and the government response to this growing unrest [5]. We noticed that both of these systems exhibited properties of predators and prey, and so we adapted the model to also simulate a generalized predator-prey system, replicating key phenomena found in the ecosystem literature such as Gause's Law, the stepped pattern of biomass accrual, and the Paradox of Enrichment [3]. With this model we also discovered some surprising limits on the Red Queen Effect: the idea that competitive populations will perform an arms race to continually outstrip the other group [4].

This endeavor is similar in scope to the work of Nicolis and Prigogine [18]. As described by [21], they were attempting to develop a rigorous theory of self-organizing behavior, and they were successful in showing that mathematical equations used to describe chemical reactions could also apply to the cyclical dynamics of a predator-prey model. However, their approach did not use a stochastic ABM method; rather, it relied on idealized equations which—though useful—are difficult for representing a diversity of agents and agent-attributes.

Our generalizability approach in [5] is most similar to that used previously by Axelrod et al. [1]. In this work, a model of political state-level alliances during World War II was successfully applied to an economics system of company-level alliances. In the political model there are five attributes—such as shared religion or border disputes—that were used as either attractors or repulsors in a pair-wise calculation of affinity across 17 countries. These affinity calculations—65,536 in total—would then determine the alliances of each country (subsequently labeled either Allies or Axis). No matter what the initial conditions, only one of two final configurations appeared each time, one of which was historically correct for all 17 countries save one.

This same model was then applied to the case of eight computer companies choosing which coalition to support between two competing versions of the UNIX operating system. This application used the same theory as that for the political model, simply adapting the attributes and relative sizes of each actor, and the model successfully predicted the real-world strategic alliances that the computer companies formed.

The primary difficulty with [1] is that there are so few agents in each system: only seventeen for the political case and eight for the business case. This limitation opens up the model to criticism, in terms of agent attributes that could, perhaps, be easily calibrated to predict a known result. Also, this system is not intended to simulate the machinations of the countries or the companies over time; rather, it merely searches for a single end state. Further, it is unclear how a set of these weights in one domain—political alliances—would help inform similar weights in another domain, such as corporate alliances.

Nevertheless, the strength of their work is that the models interactions are translatable from one domain to another, particularly regarding the underlying theory used in both cases. Such cross-disciplinary applicability is the overarching goal of the symposia and conferences that we have organized over the years, including through the AAAI, and the annual Swarmfest meetings, that are represented in this volume. Ultimately it would be more interesting and, perhaps, more useful if such trans-disciplinary models displayed similar characteristics and outputs not just at one moment in time, but over complete model runs, so that it is not just end-states that show similarities, but also the dynamics that get you there. This is a much more difficult goal to reach, of course, but perhaps also more significant and therefore more worthwhile to pursue.

2 Properties of Complex Adaptive Systems

This section looks at some of the earliest work used to formulate the paradigm of CAS and touches on the fundamental properties and key characteristics that define this paradigm.

2.1 Historical Context

In the 1960s researchers were trying to understand better the dynamics of slime mold: in particular, there was a persistent mystery in how it could transition between its active and its dormant states [15]. Biologists had long known about slime mold's strange behavior, acting as a single organism under some conditions, and devolving into individual cells under other conditions. They knew that a chemical acrasin was somehow involved, and speculated that there were pacemaker cells which would produce an acrasin and thereby attract the other cells to it. Years of study were conducted in the vain search for these pacemakers.

In the late 1960s a physicist and a mathematician (Evelyn Keller and Lee Segal) came across a paper by Alan Turing that described what he termed *morphogenesis*: the idea that organisms can form great complexity from simple roots. Published in 1952, it was one of the last papers he produced, and in it, he described a mathematical model whereby simple organisms, following just a few simple rules, could produce strikingly complex patterns [23].

Keller and Segal took the ideas in Turing's paper and developed the mathematics to describe a system of slime mold, demonstrating that it is not necessary to account for pacemaker cells in such a model. Rather, all that was required to reproduce the properties of the system were two rules: that each cell simultaneously produces, and is attracted to, an acrasin. These two simple rules were sufficient to account for the molds strange behavior, and demonstrated how this collective interaction could allow numerous individual cells to form a multi-cellular organism, one that could move about its environment and act as a single living being. A third rule, that the cells produce the acrasin under certain environmental conditions, was sufficient to explain the transition from a dormant state to an active one.

In this way, the description of a slime-mold model exhibits all the classic properties of a CAS: the agents (cells) of the slime-mold affect each other via the feedback mechanisms inherent in the two rules; they also react to the influence of the changing environment, which is sufficient to activate these two rules; once activated, the cells self-organize as an emergent property of this system; and finally, the threshold change in behavior of the slime-mold organism represents the non-linear dynamics necessary to adapt to new environmental conditions.

This re-framing of the slime-mold behavior is indicative of a systems-level approach to studying complex phenomena. This framework was also recognized as a new way to approach system-level phenomena in many other fields, such as the classic invisible hand that governs the marketplace, as found in the work of economist Adam Smith; or the contagion effect, found in social theory as well as epidemiology studies; or the study of traffic patterns and the movement of crowds. The subsequent founding of the Santa Fe Institute in 1984 by Murray Gell-Mann, a physicist; John Holland, a biologist; and others, is seen by many as the beginning of CAS as an explicit field of study [24]. They recognized the multidisciplinary nature of these phenomena, and thus brought together scholars from many different areas to begin the process of applying CAS to a wide variety of research questions.

2.2 *Complexity*

There is not yet a single, agreed-upon theory that describes complexity or a complex system equally for every situation. As with many things, it is often a matter of degree or perspective, rather than clear distinction, as to what is complex and what is not. However, we can distinguish some key characteristics of a complex system for our purposes here.

The most general distinction we use refers to Warren Weaver's division of complexity into two types: disorganized complexity and organized complexity [25]. Disorganized complexity refers to a system of many, even millions, of parts that interact at random, producing aggregate effects that can easily be described using probability and statistical methods. The example he gives is that of a very large billiard table with millions of balls rolling in different directions, colliding with each other and with the walls. Even though the path of a single ball may be erratic, or even unknown, the system itself has measurable average properties. Clearly, there is feedback in such a system: one billiard ball strikes another, and then that ball can bounce around and strike back. But this does not suffice. There is something missing in this system, without which it cannot produce self-organizing behavior.

What we are concerned with here, then, is *organized* complexity. Organized complexity refers to a system with a sizable number of agents that have correlated interactions. And since these interactions are correlated, they can produce emergent, global-level properties for the system as a whole.

> An average quantity alone is not an emergent feature. Yet statistical quantities, which define properties of an aggregation, can be regarded as simple emergent properties if they depend on a relation of the particles to each other, i.e., if they do not make sense for a single particle [12, p. 8].

Correlation among the interactions in such a system implies two things: (1) that the agents of the system exhibit feedback mechanisms; and (2) that these feedback mechanisms are, by definition, endogenous to the system itself. In this way, the agents affect each other in a correlated manner.

2.3 Agents

The term *agent* tends to be an overloaded one. Some researchers, therefore, may use an alternative, such as *particle*, to describe the individual objects of a complex system [16]. While logically sound in the way Kennedy et al. present the term, it doesn't seem to capture the autonomy, or intent, of many agents; particularly those found in social systems. Thus we use the more conventional term *agent* in our description. But we distinguish between the somewhat overlapping conceptions of agents found in CAS relative to those generally described in a Multi-agent System (MAS) [27]. CAS agents possess simple rules and attributes; are largely autonomous with only local knowledge; and, as constituent parts of a larger system, are easily replaced by similar agents without disrupting the emergent features of that system. In contrast, MAS agents tend to be more autonomous and intelligent, more complicated, and fewer in number. MAS agents also tend to fall into a strict hierarchy, whereas CAS agents are easily replaced or switched around. Contrast, for example, all the individual parts of a car with, say, a colony of bees. Each bee (CAS agent) is easily replaced with another, whereas each part of a car (MAS agent) has a strict function and placement.

Finally, *emergence* in most MAS models is usually mentioned only as something to be avoided if possible, rather than as an inherent, key property of the system. In CAS, emergence is considered a feature, not a bug.

Put another way, *building* a car is complicated. The agents are specific, diverse, and fall into a strict hierarchy. *Driving* a car is complex: dynamic and ever-changing, with multiple levels of feedback and a loose hierarchy of replaceable agents.

In our work we also consider CAS agents to be self-similar, to use a term common in the literature; i.e., the agents are largely homogeneous. It is worth noting that many published works refer to these not as homogeneous agents, but as *heterogeneous* agents, such as in Epstein [9, pp. 5–6]. We believe the discrepancy is simply a difference in emphasis. As Epstein uses the term heterogeneous, he is referring to a differentiation regarding the agent attribute-values, not the agent specifications themselves. That is, his heterogeneous agents have a range of values for their attributes, not a range of attributes. While other authors may call such agents homogeneous due to their similarity, it is useful to understand that these authors are talking about the same thing. To avoid ambiguity, we use the term *self-similar*, while also recognizing that the agents of a complex system can be different—but not too different—in terms of the rules and attributes that relate to the emergent property in question.

These differences across agents do matter, in their variety, because a particular emergent property depends upon a degree of self-similarity within the system. Consider a simple model of traffic flow as an example, with the agents as cars moving along a highway. Each agent has two rules: slow down if the car ahead is too close, and speed up if it is too far away. Under some conditions, a wave-like pattern can emerge across the ebb and flow of the cars, as one car slows, causing the next in line to slow, also. In simulations, this can occur whether the rules for slowing down and speeding up are exactly the same across all cars, or if there is some slight variation for the activation of each rule (i.e., if they are heterogeneous in attribute-values). But if some agents have rules that allow them to stop completely, or crash, or drive off the road—if they are *too* heterogeneous in their attributes—then this chaotic behavior would disrupt the emergent patterns of traffic. The system breaks down if the agents diverge too far in their rules and attributes.

Similarly, if the flocking example found in [26] were adjusted so that some agents have wildly different attributes, then *flocking* may not be a reachable state for the system. If there is no correlated feedback among the agents, then an emergent property is impossible.

The degree to which agent must be similar depends upon the characteristics of the model being studied; specifically, it depends on the emergent behavior that is of interest. For example, the agents in the traffic pattern may be made much more complex, with many more attributes, than two simple rules of when to speed up and when to slow down. Each agent's perceptions, disposition, reactive ability, and etc., could be included in the specifications. And many other agent attributes besides. But note that these attributes, and many more, only matter *to the degree that they relate to the two conditions that produce the emergent behavior*. No matter how complex the calculations that take into account perceptions, disposition, reaction times, and so forth, they ultimately determine only an expression of the two rules: when to speed up and when to slow down. The agents may be described as quite heterogeneous across all these attribute values, but they must be self-similar enough to produce an emergent traffic pattern that can be analyzed and compared to real-world data.

2.4 Agent-Level Versus System-Level Adaptation

Agent-level adaptation implies some sort of fitness function or selection criteria for agents, based on their attribute-values. This further implies some difference or capacity for change among the agents attribute-values; and more than just superficial differences, but rather functional and consequential heterogeneity. Agent-level adaptation becomes hard to distinguish under certain conditions, however. To illustrate the potential difficulty, imagine an economics model where agents sell a certain good at a certain price. The agents each have a rule that states: sell product X for no less than Y units of money. On one level, these two agents are exactly the same, in that their internal rules are the same, even if one agents current state for the value of Y is 10 dollars while another agent has his Y set to 11 dollars. The difference between the first and second agent is not the difference in rules or attributes, but in one attribute *value*. In this sense, these agents are still homogeneous, because they have the same type of rules, and they apply these rules in the same way. In another—but very real—sense, these agents are heterogeneous, adapting individually as each adjusts his price point for maximum efficiency.

This sort of change in the agents state can be termed *learning*, or *adaptation*, or even *evolution*: all words that mean essentially the same thing, but fall along an implied continuum of persistence and complexity. *Learning* is the easiest, and fastest to change, while *evolution* tends to be on longer time-scales, and is more permanent. Thus, our hypothesized economic agents may *learn* a new price-point for selling, and this price point may be updated daily. Or, some of these agents may *adapt*, changing their internal algorithms used to update this price point. Or—going even further—the agents may *evolve*, perhaps changing their modes of behavior so that they not only sell product X, but can buy it as well.

In contrast to agent-level adaptation, system-level adaptation is when a group of agents changes in a correlated way, reacting holistically to the environment. In general, we label system-level adaptation as correlated changes in the attribute-values among this group of connected agents. System-level adaptation, then, is a less substantial change in each individual agent, representing attribute-value changes rather than attribute or rule changes. And thus system-level adaptation could be represented by a flock of birds that sees a predator. The flock may shift and split apart as the individual birds try to avoid the predator, and these birds influence their neighbors to change direction as well. Even though no individual bird has changed how it reacts to seeing a predator—i.e., they haven't adapted or evolved—the flock itself can adapt to avoid the danger. It is this system-level adaptation that gives CAS its power: collectives reacting intelligently to the environment, with complex dynamics and versatility, even though they are comprised of simple agents with simple rules.

2.5 Feedbacks

Feedback, simply defined, means that the outputs of a system at time t affect the inputs of that system at time $t+1$. As the agents in a complex system interact, the results of some interactions may influence future interactions. It is this influence that represents the feedback within the system itself. In the previously mentioned model of traffic patterns along a highway, one car that slows down in response to the car in front of it may then produce a similar effect in the next car in line. This action/response that can easily produce a wave of congestion along the highway is due to feedback between the cars, from one to the next in line. It is worth pointing out that the term *wave* is apt in this case, as it describes a pattern of behavior across multiple agents, much like a wave in the ocean, even though the agents participating in the pattern change over time. This matches well with how Holland and others have described emergence in complex systems:

> Emergent phenomena in generated systems are, typically, persistent patterns with changing components [13, p. 225].

Note also the distinction between this organized feedback as compared to the disorganized complexity of our billiard table. While it is true that one collision between two balls alters the course of future collisions, it does not affect the course of future collisions *persistently*; that is, if one colliding ball happens to bounce to the north, it does not mean that the next ball struck will also bounce northward.

> Relationships in these systems are mutual: you influence your neighbors, and your neighbors influence you. All emergent systems are built out of this kind of feedback [15, p. 120].

The key point here is that such reciprocal influence among neighbors is more significant when it creates measurable, global properties. The action/reaction patterns represent the correlations within the system that make up these global properties. While our traffic pattern example may have measurable statistical properties—such as how many cars traverse the highway in a given day—these measurements do not fully capture the wave-like behavior of the system. It is by identifying the correlated feedback that we find a richer, and therefore more interesting, description of the system.

2.6 Endogenous Versus Exogenous Factors

One may want to consider the first action that sets the pattern in motion—is it an endogenous or exogenous instigator? While the resultant pattern is certainly endogenous to the system, the initiation of that pattern may be either. It can sometimes be difficult to characterize effects as one or the other, and how the system itself is defined may further confuse the distinction. However, by defining correlated feedback as a key property of a CAS, we bypass this argument in favor of defining what the feedback represents, and what it tells us about the system.

If an external effect sets off a chain reaction of persistent patterns, then the underlying properties that allow this chain reaction to occur are of distinct interest for understanding the system. If, however, there is persistent and recognizable feedback that comes from outside of the system, then we consider this feedback to be significant regarding our understanding of the system properties. Therefore, when we define a system, we use the method and type of feedback as a key attribute.

Consider the example of a marketplace. Such a system may encompass agents that buy and sell products, or stock in companies; it may include the concept of wealth, earnings, inflation, etc.; and it may also be affected by regulatory bodies, such as the Federal Reserve. If one defines the system as only the agents and how they interact with each other, then the actions of a Federal Reserve would be exogenous to this system. However, these actions by the Federal Reserve—whatever they may be—are clearly influenced by the state of the market. Furthermore, they are likewise designed to influence the future state of that market. This is a significant level of feedback that should be accounted for when studying the system, i.e., the market.

Another way of determining whether certain factors are exogenous or endogenous to the system is to consider whether or not the feedback goes both ways: the agents affect the environment even while the environment affects the agents. This is distinct from a model of, say, an ecology which has sunlight as an external factor. The sun cycles through day and night, as well as annual cycles of summer and winter, and these cycles generally affect the behavior of most ecological systems. But the agents in this system cannot truly affect the behavior of the sun. While defining what encompasses a system, and what potential factors are internal or external to that system, it is more important to note the level of feedback that exists between those factors, as this is both definitional and functional to the system being studied.

2.7 Emergence and Self-Organization

The term emergence, like complexity, has not yet reached a consensus definition. Some researchers distinguish between weak emergence and strong emergence, and use this definition as representing a fundamental law.

> If there are phenomena that are *strongly emergent* [emphasis added] with respect to the domain of physics, then our conception of nature needs to be expanded to accommodate them. That is, if there are phenomena whose existence is not deducible from the facts about the exact distribution of particles and fields throughout space and [time] (along with the laws of physics), then this suggests that new fundamental laws of nature are needed to explain these phenomena [6, p. 1].

This idea would seem to indicate that a strongly emergent property is similar to the idea of gravity: gravity is a fundamental law, a property of matter; but gravity is only apparent as one particle relates to another. In this view, it is not that the rule cannot be modeled by the agent, but rather it cannot be understood except in terms of other agents.

In our definition of emergent behavior, we adopt this idea of relations among agents in the system, as in the way we have previously defined correlated interactions. A traffic pattern cannot really exist with only one car, and a colony of ants cannot be a colony if there is only one ant. In this way, emergent behavior is a property of a system that is at a different scale than the parts of the system [19]. In a similar vein, emergence is the macro-level behavior that is not defined at the macro-level, but rather depends upon the rules and interactions of agents defined at the micro-level.

Consider a few examples of typical emergent behavior. There are the cars as agents, in the example cited previously. There is also the example of bees or ants, following simple rules to forge for food or build a nest. Johnson talks at length about the city of Manchester, England, during the 19th century [15]. He uses it to illustrate how a city with tens of thousands of people, yet absolutely no central planning, still managed to organize itself in distinct patterns, such as areas of the working class separate from the nicer middle-class neighborhoods.

> The city is complex because it has a coherent personality, a personality that self-organizes out of millions of individual decisions, a global order built out of local interactions [15, p. 39].

The brain is also often cited as a complex, adaptive system, with intelligence (or even some sub-set of intelligence, such as vision) as an emergent feature. In our CAS models, we look at a number of emergent features, such as the self-organization of the agents and the aggregate behavior of the system [3–5].

The *self* in *self-organization* refers to the state of an individual agent in a complex system. This agent follows its own local rules, and uses its own own attributes in applying those rules. Let us consider a simple model of an ant colony. For the purposes of illustration, this model need not be realistic. Assume each individual ant has the same three rules: (1) search randomly across the environment for food; (2) if you find food, carry it back to the colony and leave a scent trail; (3) if you find a scent trail, follow it until you find food.

If one ant finds food, then this new attribute—"I have food"—activates the rule to carry a piece of the food back to the colony and leave a scent trail. Now, by leaving the scent trail, this ant can affect the current state of any other ant that happens upon that trail. A new ant, finding the scent trail, will activate its own rule to follow that trail to the food source, at which point it will also carry a piece back to the colony, and add to the trail. In this way, a significant subset of the ant colony organizes itself to systematically collect the food and bring it back to the colony. The individual agents—in this case, the ants—are acting with limited knowledge and simple rules. But by providing feedback to other agents, and influencing them to act in similar ways, they produce the correlations of behavior that represent the organization of the overall system; i.e., the self-organization that emerges from these interactions, defining the local increase in complexity.

2.8 Natural Biases of Complex Systems

The framework of CAS directly challenges two distinct biases that tend to affect our understanding of the agents in a complex system: (1) a hierarchical bias; and (2) a complexity bias. A hierarchical bias can be illustrated by the tendency to view a complex system in terms of a leader directing the activities of all the other agents. As Johnson points out, colonies of ants have previously been viewed as the queen controlling the colony as a whole; however, this fails to capture the amount of autonomy present among the other ants [15]. And, with a little reflection, it becomes obvious that a queen ant simply would not have the bandwidth necessary to communicate to all the other ants, and direct them in their daily tasks. Fundamentals of information theory demonstrate that such would be impossible. Most ants do not come into contact with the queen, and they do not have much to say when they do. Only a few things can be communicated via pheromones that ants exchange, and complicated task lists are not among them.

In much the same way, the growth of Manchester previously mentioned, and the distinctions that emerged between, say, rich and poor neighborhoods, was deeply surprising to those who thought that such patterns of growth could only be achieved by directed action, through some sort of governing body.

These strange phenomena—global properties of systems, such as those represented in the growth of Manchester, or in Smith's invisible hand theory—did not go unnoticed or unstudied. However, as with the peculiar behavior of the slime mold, researchers struggled to frame a model that could explain these global effects using a hierarchy that they intuitively felt must exist. The development of CAS tools and models, therefore, represent a new methodology to remedy the shortcomings of previous methods. We no longer have to assume that the behavior is directed in a hierarchical fashion. Distributed intelligence and decision-making does not require a central governing authority. Correlated feedbacks among autonomous agents are enough to describe and model these behaviors.

CAS methods of analysis also help resist the *complexity* bias for hypothesized agents that is often found when studying complex systems. This is closely related to the hierarchical bias, in that leader-agents are assumed to be more complex, to account for the level of control needed in a leadership model. In other words, the leader must be smarter: more capable and more complicated. Also, the required network among such agents would be, by necessity, more complicated and long-reaching, to allow for instructions to be passed to each agent in the system. If one is to assume a hierarchical system in, say, an ant colony, then the modeler must answer the question: how are orders conveyed to each worker ant? The consequences of a complexity bias is a more unwieldy, computationally expensive, and fragile model.

A CAS is inherently simpler. Each ant does not need instructions; rather, they can be programmed with just a few rules of behavior. In such a model, the ants do not even have to be aware of the state of the colony as a whole; they only need to know their own current state and apply that information to their current environment. Similarly, a slime-mold model doesn't require a complex pacemaker cell if a simpler CAS model is able to replicate the organism's complex behavior without it.

This release from both the hierarchical bias and complexity bias in the agent-level description of a system is more satisfying, as it follows Occam's Razor: the simplest explanation for a phenomenon is the preferred one. And the beauty of this paradigm is also found in the fact that the simpler explanation—the emergent, distributed explanation—is also less expensive to implement, both computationally and in the real world. Fitness functions that are inherent in nature are always pushing the system, any system, toward more efficient use of resources. And thousands of autonomous, simple ants that don't require constant instruction are surely more efficient—and more robust—than a model that has one central, complicated, irreplaceable, and over-worked queen.

3 Conclusions

The assumptions inherent in Complex Adaptive Systems have allowed us to more productively study challenging and complex phenomena, in both nature and society. It has allowed us to uncover intriguing similarities in domains that are seemingly far apart. And it allows us to focus on the agent primitives in our models, as direct analogues to real-world behavior. This inherent transparency is a key benefit of

Agent-based Modeling, as an antidote to black box simulations that may be correct, but are, by definition, obscure and hard to analyze.

By using the CAS paradigm, with correlated feedbacks among simpler agents, the features of the system of interest are allowed to emerge from dynamic agent interactions, as they do in the real world, rather than be dictated in a top-down, complicated manner. Ultimately, this gives our models the inherent flexibility needed to simulate systems even when conditions or outcomes are different than expected. This flexibility makes our models more fundamentally robust: able to adapt to a variety of environments, only some of which may be anticipated. Thus, the CAS approach, and its inherent flexibility and distributed robustness, creates models that can capture results we did not already expect to see, leading some to call CAS: The Science of Surprise.

Acknowledgements Some of this material appeared previously, in a slightly changed form, in Managing Complexity: Practical Considerations in the Development and Application of ABMs to Contemporary Policy Challenges [2].

References

1. Axelrod, R.: Agent-based modeling as a bridge between disciplines. In: Judd, K.L., Tesfatsion, L. (eds.) Handbook of Computational Economics, vol. 2, pp. 1565–1584. Elsevier, Amsterdam (2006)
2. Carmichael, T.: An overview of agent based models. In: Hadzikadic, M., OBrien, S., Khouja, M. (eds.) Managing Complexity: Practical Considerations in the Development and Application of ABMs to Contemporary Policy Challenges, vol. 504. Springer, Berlin (2013)
3. Carmichael, T., Hadzikadic, M.: Emergent features in a general food web simulation: Lotka–Volterra, gause's law, and the paradox of enrichment. Adv. Complex Syst. **16**(08), 1350014 (2013)
4. Carmichael, T., Hadzikadic, M.: Predator-prey dynamics and the red queen hypothesis: putting limits on the evolutionary arms race. J. Policy Complex Syst. **2**, 1 (2015)
5. Carmichael, T., Hadzikadic, M., Dréau, D., Whitmeyer, J.: Towards a general tool for studying threshold effects across diverse domains. In: Ras, Z.W., Ribarsky, W. (eds.) Advances in Information and Intelligent Systems, pp. 41–62. Springer, Berlin (2009)
6. Chalmers, D.J.: Varieties of emergence. Technical Report preprint, Department of Philosophy, University of Arizona, USA (2002)
7. de Vries, H., Biesmeijer, J.C.: Modelling collective foraging by means of individual behaviour rules in honey-bees. Behav. Ecol. Sociobiol. **44**(2), 109–124 (1998)
8. de Vries, H., Biesmeijer, J.C.: Self-organization in collective honeybee foraging: emergence of symmetry breaking, cross inhibition and equal harvest-rate distribution. Behav. Ecol. Sociobiol. **51**(6), 557–569 (2002)
9. Epstein, J.M.: Generative Social Science: Studies in Agent-Based Computational Modeling. Princeton University Press, Princeton (2006)
10. Epstein, J.M.: Remarks on the foundations of agent-based generative social science. Handb Comput Econ **2**, 1585–1604 (2006)
11. Erol, K., Levy, R., Wentworth, J.: Application of agent technology to traffic simulation. In: Proceedings of the Complex Systems, Intelligent Systems and Interfaces (1998)
12. Fromm, J.: Types and forms of emergence. arXiv:nlin/0506028 (2005)
13. Holland, J.H.: Emergence: From Chaos to Order. Addison-Wesley, Boston (1998)

14. Johnson, N.F.: Two's Company, Three is Complexity: A Simple Guide to the Science of all Sciences. Oneworld Pubns Ltd, London (2007)
15. Johnson, S.: The Connected Lives of Ants, Brains, Cities and Software. Scribner, New York (2001)
16. Kennedy, J., Eberhart, R.C., Shi, Y.: Swarm Intelligence. Elsevier, Amsterdam (2001)
17. Midgley, D., Marks, R., Kunchamwar, D.: Building and assurance of agent-based models: an example and challenge to the field. J. Bus. Res. **60**(8), 884–893 (2007)
18. Nicolis, G., Prigogine, I.: Self-Organization in Nonequilibrium Systems (1977)
19. Ryan, A.J.: Emergence is coupled to scope, not level. Complexity **13**(2), 67–77 (2007)
20. Schelling, T.C.: Dynamic models of segregation. J. Math. Sociol. **1**(2), 143–186 (1971)
21. Sumpter, D.J.: The principles of collective animal behaviour. Philos. Trans. R. Soc. B Biol. Sci. **361**(1465), 5–22 (2005)
22. Tay, N.S., Lusch, R.F.: A preliminary test of hunt's general theory of competition: using artificial adaptive agents to study complex and ill-defined environments. J. Bus. Res. **58**(9), 1155–1168 (2005)
23. Turing, A.M.: The chemical basis of morphogenesis. Philos. Trans. R. Soc. B Biol. Sci. **237**(641), 37–72 (1952)
24. Waldrop, M.M.: Complexity: The Emerging Science at the Edge of Order and Chaos. Simon and Schuster, New York (1993)
25. Weaver, W.: Science and complexity. Am. Sci. **36**(4), 536–544 (1948)
26. Wilensky, U.: Netlogo flocking model. Northwestern University, Evanston, IL, Center for Connected Learning and Computer-Based Modeling (1998)
27. Wooldridge, M.: An Introduction to Multiagent Systems. Wiley, New York (2002)

A Cognitive-Consistency Based Model of Population Wide Attitude Change

Kiran Lakkaraju and Ann Speed

Abstract Attitudes are an important component in human decision making. We focus on developing a model of attitude diffusion across people that includes social and cognitive factors. We will describe the model and the underlying theories that support its elements. Our results focus on better understanding the resiliency of attitudes—how long does it take for attitudes to change and what do they change to?

1 Introduction

Attitudes are "general and relatively enduring evaluative responses to objects" where objects can be "a person, a group, an issue or a concept" [40]. Attitudes are shown to have an impact on, and can sometimes predict, the behaviors of individuals (e.g., voting behavior [15], consumer purchases [9, 14]).

Understanding population wide attitude change is thus an important step to understanding the behavior of societies. For instance, consider the change in attitudes towards global warming and the environment that has resulted in a significant change in public policy and national priorities [4].

While there are a number of factors that influence attitude change [41], we focus on two. The first is social—individuals are exposed to various attitudes and information through interaction with others. Family, friends, acquaintances, and the media all influence the attitudes of individuals by providing new information/opinions. Reference [42] demonstrated that the social context of an individual can impact a large number of aspects of that person.

The second factor is cognitive—individuals tend to hold a set of attitudes that are consistent with each other [27, 31, 33]. According to these *cognitive consistency theories*, an individual holding a strong positive attitude towards environmentalism

K. Lakkaraju (✉)
Sandia National Laboratories, Livermore, CA 94550, USA
e-mail: klakkar@sandia.gov

A. Speed
Sandia National Laboratories, Albuquerque, NM 87185-1327, USA
e-mail: aespeed@sandia.gov

T. Carmichael et al. (eds.), *Complex Adaptive Systems*, Understanding Complex Systems, https://doi.org/10.1007/978-3-030-20309-2_2

should also hold a strong positive attitude towards recycling; if they do not, the attitudes are inconsistent with each other and could cause an uncomfortable feeling (i.e. *cognitive dissonance*) which tends to result in either attitude or behavior change [41].

The perspective of the authors is as follows, we argue that the phenomena of attitude change can be reasonably captured through a multi-scale *complex adaptive system* beginning with two levels: (1) a social level where autonomous agents representing individuals interact and exchange information; and (2) a cognitive level, where individual concepts interact in order to preserve consistency.

These two factors are tightly linked. The social context of an individual can influence the information they are exposed to as well as how amenable to change the individuals are [42]; while the individual's existing attitudes may distort or diminish the impact of information [11, 33] and could impact the composition of an individuals social network.

Our research aim is to develop a model that can capture these two levels and the interaction between them. This paper presents preliminary results from the development of a new computational model of attitude change that incorporates both social and cognitive factors. The model represents individuals as networks of concepts (e.g., beliefs) that interact with each other. Social structure is implemented through a social network that limits the interaction between agents.

The dynamics of this model are explored through a simple attitude change experiment. The objective is to understand the *resiliency* of attitudes—how long does it take for attitudes to change and what do they change to? Two populations of agents with different attitude structures interact. We imagine this as a new population of individuals being introduced into an existing population. We show that the population does reach agreement (all agents have similar attitudes), and that the state that is agreed upon is a function of the initial distribution of attitudes in the population. In addition, we track time to agreement and show that it is a function of the social network and the initial distribution of attitudes in the population.

We first describe some of the empirical findings on attitude change and consistency and summarize some of the key phenomena and factors that can influence attitude change. We then describe the parallel constraint satisfaction model which we will use to implement cognitive networks. Finally, we describe our model and illustrate some properties through experiments on a simple attitude change scenario.

2 Attitudes and Attitude Change

Research on attitudes and attitude formation indicates that an attitude is generally considered an opinion or a feeling about something that has both a valence (i.e., positive or negative) and a strength [44]. These opinions influence behaviors and influence how new information is received and processed [7, 8, 28, 33, 44].

Even though research has traditionally focused on attitudes as singular feelings (e.g., one either has a positive or a negative attitude about American football), research in the last decade has highlighted the fact that an individual can hold two attitudes about the same object, and that some of these attitudes are explicit and some are implicit [28, 44]. Further, research has shown that such dual attitudes can even conflict with one another. For example, by explicit measures (e.g., questionnaires), a Spaniard might express his love of American football, but when measured using implicit measures, he may reveal some disdain for the sport, and this may influence nonverbal behaviors or motivations [7]. These are in stark contrast to ambivalence (e.g., positive feelings for the flavor of a food but negative feelings about the calorie content of that food).

As an example, in five experiments, Rydell and McConnell 2006 [28] demonstrated the ability to create separate implicit and explicit attitudes about a fictional character, Bob, and that different manipulations impact the explicit and implicit attitudes differently. What is particularly interesting about this series of experiments is that the authors do not rely on previously held attitudes they create attitudes in the laboratory. However, studies of attitudes using pre-existing beliefs tend to yield similar results [8, 10, 11].

These findings are important because of the implications for behavior. Research has demonstrated that explicit, controllable behaviors change as explicit attitudes change. Research has also demonstrated that implicit attitudes change more slowly than do explicit attitudes, and that implicit attitudes predict a separate set of behaviors than do explicit attitudes [7–9, 22, 28]. Specifically, explicit attitudes predict behaviors over which the individual holding the attitude has control. Implicit attitudes tend to predict behaviors that are more subtle (e.g., nonverbal behaviors) and other behaviors over which the individual holding the attitude has little or no control.

If the findings in [28] about the factors that cause attitude change hold (e.g., the effect on attitude change of the relative proportion of negatively versus positively valenced evidence), that has implications for the kinds of things that need to be considered in efforts to change attitudes including specific actions to be taken as well as the amount of information needed to enact change in both explicit and implicit attitudes. Otherwise, the risk is that behaviors that are under conscious control may come to reflect the explicit attitude while behaviors that are more subtle, but possibly as powerful, will reflect the contrary implicit attitude.

Another variable that seems to impact explicit attitude change is the effect of cognitive load. Thus, current research suggests two routes to attitude change. In the systematic route individuals are assumed to expend deliberate effort to evaluate a message; including evaluation of the source of the message, the validity of the arguments, and relating the elements to their own attitudes. However, in situations of high cognitive load, that is periods where individuals are distracted and cannot expend cognitive effort, the peripheral route is taken. In these instances individuals do not evaluate the argument but rather use heuristics to decide whether to change their attitude. These heuristics can depend upon the message structure (if the message has many arguments individuals are more likely to be persuaded); or based on the context of the message (if many people already agree to the message the receiver is more likely to agree). Current work shows that this dual process model accounts for instances of attitude change that the systematic model does not [41].

2.1 Attitude Change and Cognitive Consistency

One important question to ask is—when might an attitude change? The complimentary question is—when might the thinker distort new information in order to fit with an existing attitude or opinion? Several such factors have been mentioned above, but it is worthwhile to discuss those in greater detail. Generally, attitudes are less likely to change in an enduring fashion when (1) they are strongly held, (2) the thinker has less time or fewer cognitive resources to think through the logic of an argument that would result in attitude change, and (3) if they are implicitly, rather than explicitly held beliefs [11, 28, 44]. In terms of explicit attitudes, when the attitude isn't changed by incoming information, the thinker often distorts that incoming information to fit with his or her current attitude. Overall, this process of either changing ones attitude or biasing incoming information seems to be mediated by a drive for consistency between beliefs, attitudes and action [18, 25, 27, 33].

According to these cognitive consistency theories, individuals strive to hold a set of attitudes that are consistent with each other. For example, an individual holding a strong positive attitude towards environmentalism should also actively engage in recycling or composting household waste; if they do not, the attitudes are inconsistent with each other and could cause an uncomfortable feeling (i.e. cognitive dissonance) which tends to result in either attitude or behavior change [41]. The important idea is that attitudes interact with each other; a change in one attitude can have an effect on another.

Some of the classic studies on consistency demonstrate that people have a tendency to rate items differently when they occur in isolation than they do when they occur within an attitude-formation context.

In a subsequent series of studies, [33] demonstrated that attitudes about situations tend to form early in the evidence accrual process and that these forming attitudes tend to bias subsequent information, strengthening the original attitude. These authors also demonstrated that if the subject's attitudes later changed, they then biased information in the opposite direction. That is, if Information A is contradictory to a subject's initial attitude, he will discount it. However, if that same subject changes his attitude in the end, the discount on Information A will no longer be apparent. Rather, he will have discounted Information B, which was supportive of his initial attitude.

Specifically, the authors presented two sets of questions to subjects, one that involved a series of unrelated vignettes, and the other that consisted of questions all relating to a single, more extended story line. Importantly, the two sets of questions were otherwise identical. Thus, the authors measured the change in attitudes that was attributable to the subjects forming a coherent opinion in the extended story relative to the answers they gave in the unrelated vignettes. Over four studies, they found that people discount information that is contrary to their forming opinion, even before they have to officially declare that opinion. They also demonstrated that when people changed their attitude (i.e., they leaned one way initially and changed their attitude to the opposite opinion in the end), their evaluation of the evidence also changed to maintain coherence with their eventual decision. Thus, pieces of information that were weighted highly initially were down-weighted in the end if they did not support the opposite view.

Additional evidence for the drive for consistency is demonstrated in [27]. These authors utilized a number of explicit and implicit measures to demonstrate that a tendency toward consistency is, indeed, behind information distortion, rather than other processing goals such as reduction in cognitive effort or as a method for increasing the distinctiveness of the alternatives under consideration.

Interestingly, one social factor that seems to influence the malleability of a given individual's attitudes is the heterogeneity of their social network with regards to that attitude. In a series of four experiments 2 in the laboratory and 2 using existing, real-world social networks [19] demonstrated that the more heterogeneous a given individual's network is with regard to an issue (e.g., attitudes about capital punishment), the more ambivalent the person is in their attitude, and the less certain they are of their attitude. Similarly, the more homogeneous an individual's social network is regarding a particular issue (i.e., the more attitudinally congruent an individual's network), the stronger that individual's attitudes are, and the less open that individual is to attitude change. Interestingly, these researchers also found that when subjects in the laboratory were exposed to attitudinally congruent versus heterogeneous sets of opinions about the same issues, only not in the context of a social network, subjects in the attitudinally congruent condition demonstrated increased attitude change relative to those in the attitudinally heterogeneous condition when exposed to a persuasive article about the topic of focus. Although this effect was not statistically significant, the reversal of trends is curious and warrants further investigation.

In a series of studies on information distortion, [27] manipulated both implicit goal priming and explicit reporting of goals used. They demonstrated that the only process goal that drove the evaluation of evidence in progressive choice tasks was consistency. The other goals tested in these experiments included the goal of reducing the effort necessary for evaluating incoming information and the goal of increasing the separation between alternatives.

In terms of attitude change, possibly the most well-known demonstration of attitude change in order to achieve consistency is the extensive literature on cognitive dissonance. In many experiments, Festinger and colleagues demonstrated the ability to get participants to express opinions that corresponded to their behavior, even when it was clear that their original attitude was quite the opposite [41]. One example of a typical dissonance study has participants engage in an extremely boring experimental task. Then subjects are asked to tell incoming participants about the task. When subjects are asked how they personally felt about the study, those who told future participants it was fun and engaging were more likely to rate it as more enjoyable than those who did not. Interestingly, if subjects can attribute their incongruent behavior to some external factor, such as receiving a large sum of money for their fib, or being instructed by the experimenter to fib, they are less likely to report that they personally felt the task was fun.

Another variant on the consistency theme was demonstrated by [28]. These researchers demonstrated that explicit attitudes can change fairly rapidly in the face of counterattitudinal information. They also demonstrated that the amount of attitude change interacted with the homogeneity of the information that was considered in the development of the initial attitude. Specifically, in one of a series of 5 experiments concerning a fictional character, Bob, subjects learned about Bob's character through reading and responding to 100 facts. For some subjects, all 100 facts about Bob were consistent (he's a nice guy.) For the other subjects, 75% were consistent and 25% were counterattitudinal. When subjects then received additional information about Bob, those who were in the 100% condition, who received counterattitudinal information, displayed a more dramatic shift in explicit attitudes about Bob than did those in the 75% condition.

3 Summary of Factors that Influence Attitude Change

Table 1 summarizes four phenomena that we argue influence attitude change. Table 2 summarizes factors that parameterize the four phenomena. We believe that future models of population-wide attitude change should implement the four processes parameterized by the factors. In the next section we will develop one such conceptual model. Our research aim is to implement the model and manipulate the factors to understand attitude change.

While the preceding review begins to demonstrate a fairly complex picture of attitude change and information distortion, and while there are even more variables that seem to influence these effects [41, 44], for the current model, we chose to

Table 1 A summary of important attitude change phenomena

Phenomena	Description and its effect
The drive for consistency	Maintaining a consistent set of attitudes can drive attitude change as well as information distortion
Information distortion	One way in which to maintain consistency between attitudes—instead of changing an attitude one can decrease the importance or impact of incoming information
Persuasion route	Peripheral and systematic routes determine which aspects play a role in attitude structure; systematic routes focus on logical arguments whereas peripheral routes utilize heuristics. The persuasion route also influences the durability of attitude change
Implicit/explicit attitudes	These two types of attitudes influence different behavior. Implicit attitudes affect spontaneous behaviors while explicit attitudes affect deliberate behaviors. Their change profile is also different, with explicit attitudes tending to be more malleable than implicit attitudes

Table 2 Important factors that influence the phenomena in Table 1

Factor	Effect
Cognitive load	Impacts persuasion route
Attitude relationships	Related attitudes influence each other through the drive for consistency
Social network	The structure of relationships between individuals influences the rate of information diffusion; in addition the distribution of attitudes of an individuals acquaintances can impact the strength of an attitude

focus specifically on factors that might influence the change of explicit attitudes. As such, we have chosen some very particular variables to manipulate in the simulation, rather than trying to capture all of the nuances presented in the vast attitude change literature (see e.g. [41]). Two aspects are of concern to us. The first is the idea of *bidirectionality* in inference. As seen in [33] the notion that individuals gather information, then come to a conclusion is not always true. Rather, it seems that individuals are constantly integrating information into their growing conclusion, and that preliminary conclusions impact how one integrates and perceives information. Bidirectionality is a special case of the idea of *wholistic processing*, that all concepts influence and are influenced by each other [26].

The second important point is that there is a rich set of behavior based on the strength of an attitude. As [28] shows, it is important to capture how the strength and directionality of an attitude as it can impact how that attitude changes. Using a binary approximation (either one believes in X or does not believe in X) doesn't capture this rich set of behavior.

4 Consistency and Parallel Constraint Satisfaction

The drive for cognitive consistency has been discussed since the 1950's in a variety of contexts. Theories such as cognitive dissonance and balance theory utilize a notion of consistency. It is only recently through the use of *Parallel Constraint Satisfaction* (PCS) models, however, that computational models have been developed that allow a more complete exploration of these processes.

A PCS model is a type of connectionist, attractor neural network. In this model there are a set of variables that can take on a value from a specified range. The variables have weighted links to other variables, forming a *constraint graph*. Each variable is influenced by and influences its neighbors in the constraint graph.

Depending upon the domain that the PCS model is applied to the variables, values of the variables, and links represent different things. When studying impression formation, the variables represent stereotypes and traits, values represent belief, and links represent associations between stereotypes and traits [16]. Since our objective is the study of attitude change, we view variables as representing concepts, hypotheses or information, such as the concept of "environmentalism", the hypothesis of anthropogenic global warming, or information such as data on temperature fluctuations. Henceforth, we refer to variables as "concepts". Loosely following attitude change literature, we describe the value of a concept by its *valence*—indicating the sign and its *strength* indicating its magnitude. An attitude towards a concept is then represented by the valence and strength of the concept. Positive valences indicate support of the concept and negative valences indicate non-support of the concept. To simplify discussion, we refer to the *value* of a concept as the combination of valence and strength. In our model values are continuous and range from -1 to $+1$.

Weighted links between concepts indicate the strength of influence between the concepts. A positively weighted link indicates two concepts that positively influence each other—as one of the concepts increases in value, the other concept increases as well. For example, we would posit the concept of anthropogenic global warming to have a positive link to the concept of recycling. The more one believes humans caused global warming, the more one should believe in reducing human impact on the environment.

Negatively valued links indicate concepts that do not agree with each other; as one of the concepts increases in value, the other will decrease. For example, the more one believes humans have caused global warming the less one should believe that natural climate cycles are the cause of global warming.

Links between concepts can be viewed as constraints between the values of the different concepts. Two concepts with the same valence but a negative link between them would be considered inconsistent with each other. Two concepts with differing valences but a positive link between them would also be considered inconsistent. We can measure the consistency of a set of concepts by calculating what is called the *energy* of the system [13, 26]:

$$E = -\sum_i \sum_j w_{ij} a_i(t) a_j(t) \tag{1}$$

where w_{ij} is the weight of the link between concepts i and j; $a_x(t)$ is the valence of concept x at time t. The energy of the system decreases if the concepts satisfy the constraints between them and increases if the concepts do not satisfy the constraints. Because of this property we can view the energy of the system as a measure of the consistency of the set of concepts, the lower the energy the higher the consistency.

PCS models utilize a connectionist approach to find a consistent set of concepts: the value of each concept is updated according to a non-linear *activation function*. One activation function that is often used is [16, 36]:

$$\begin{aligned} a_j(t+1) &= \\ a_j(t)(1-d) &+ \begin{cases} net_j(\max - a_j(t)) \text{ if } net_j > 0 \\ net_j(a_j(t) - \min) \text{ if } net_j \leq 0 \end{cases} \end{aligned} \tag{2}$$

where:

$$net_j = \sum_i w_{ij} a_i(t) \quad (3)$$

and d is a decay term that is set to 0.05.

The update rule modifies the valence of a concept based on the valences of other concepts. net_j captures the input to a concept j by summing the valences of all nodes against their weight to j. If a concept does not influence another concept the weight is set to 0. max and min are the maximum and minimum values that a concept can take.

Each agent synchronously updates all concepts in the network and repeats this process until no significant change[1] has occurred between iterations. We call the repeated application of the update rule until no significant change has occurred the *update process*.

Previous work on asynchronous update protocols have shown that an update rule like Eq. 3 will result in a state that satisfies the constraints [13]. The result of the synchronous update rule has not been proven in the same manner, however empirical results indicate that this update rule does reach a state that satisfies the constraints.

Through the use of synchronous updating and establishing constraints between concepts, PCS models capture the important properties listed above. Value changes in concepts cascade to other concepts as a function of link strength which is one aspect of wholistic processing. The use of a continuous range for concept values allows the modeling of attitude strength.

4.1 An Example of Constraint Satisfaction

The left hand side of Fig. 1 shows an inconsistent PCS network with 5 concepts, represented as circles. The value is indicated within the node and can range from -1 to $+1$. Link weights are shown adjacent to the link and also range between -1 and $+1$.

Two pairs of concepts are inconsistent with each other, Concepts 3 and 4 have a positive link but have opposite valences; and Concepts 3 and 5 have a negative link but have the same valence. The energy of the system, calculated according to Eq. 1 is 0.005. After executing the update process we have the network on the right, which has an energy of -0.029. In the updated network, the inconsistencies have been resolved, concept 4 now has a positive valence and concept 5 now has a negative valence.

Note that the decay parameter can have an effect on the final valences of the concepts—a decay value closer to 0.0 will result in less change of the concepts.

[1]Defined as a maximum of .01 difference between concept valences between iterations.

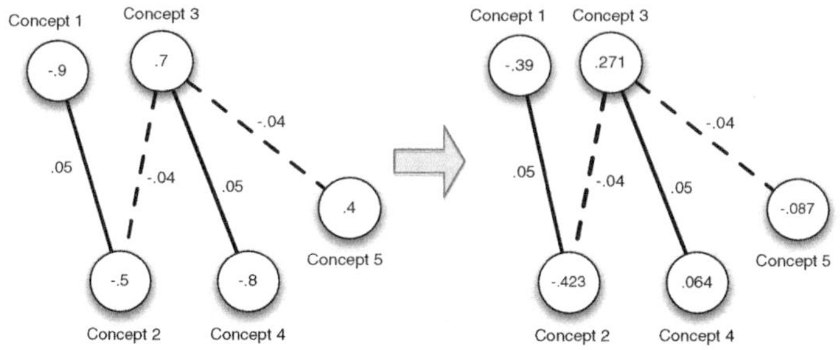

Fig. 1 Example of the updating procedure in a PCS network

5 A Socio-cognitive Constraint Satisfaction Model of Attitude Change

We propose a model of attitude change that brings together both social and cognitive factors that influence such change. Our model is composed of three elements:

$$< A, G, C > . \tag{4}$$

A represents a set of n agents that represent individuals. C is a *cognitive network* that represents the internal cognition that drives change in an attitude. We use the PCS model defined above for the cognitive networks of the agents. Each agent has the same number of concepts (m), and the same weights between links (W), however each agent can differ in the strengths and valences of those concepts.

The social network, G, is a graph of size n that defines the social structure of the population. Agents are restricted to interacting only with their neighbors.

We assume turn based dynamics—at each timestep the following actions are taken:

1. A single agent is uniformly randomly chosen from the population. We call this agent the *speaker*. One randomly chosen neighbor of the speaker is designated the *hearer*.
2. A single concept is chosen as the *topic* of communication.
3. The speaker communicates the valence and strength of the topic in its cognitive network to the hearer. This is the *communication* step.
4. The hearer changes the value of the topic (capturing the idea of integrating a speakers beliefs). This is the *information integration* step.
5. The hearer modifies the strengths and valences in its cognitive network in order to find a more consistent set of beliefs. This is the *hearer update* step.

Figure 2 illustrates one turn, progressing from left to right. First a speaker and a hearer are chosen from the social network. Then one of the concepts is chosen as a topic (denoted by the label "topic" in the example). All agents have the same set of concepts so it is guaranteed that the hearer and speaker will have the topic. However,

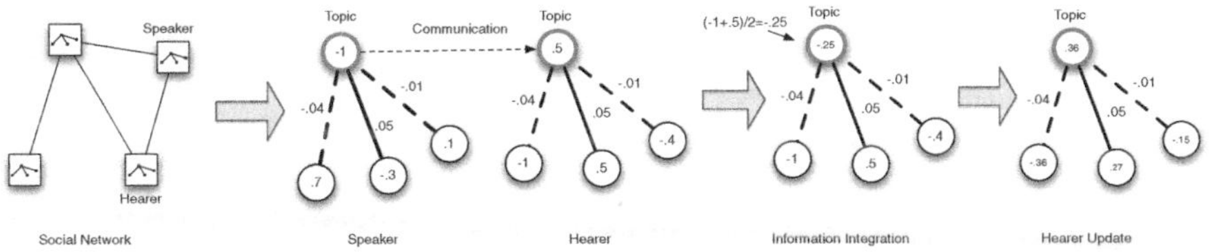

Fig. 2 Illustration of the dynamics of one timestep in the simulation

they might differ in value; we can see that the speaker strongly disbelieves the concept (value is -1) while the hearer moderately supports the concept (value is 0.5).

The communication step represents interaction and information exchange between agents. This is, of course, a highly simplified version of what occurs in real human dynamics. For this model, we assume that agents honestly communicate their values with each other.

Information integration is the process by which an agent changes its beliefs in response to the received message. Motivated by the work of [28] our information integration process depends upon the strengths and valences of the speaker and hearer. The larger the difference between the valences, the larger the change in the hearer's valence. Specifically, we take the average of the two valences in the instantiation of the model. After the information integration step, the hearer's valence has changed to -0.25 as it is the average of -1 and $+0.5$.

Finally, in the last step the hearer updates the valences of the concepts by running the update process. In the example, we can see that the information integration step has resulted in an inconsistency between the topic and another concept. The update process resolves this consistency, however the values of the other concepts have been altered as well. Even though consistency has been achieved, and the hearer has, essentially, the same attitude as before, there has been some change in response to the communication. It is this ability to capture the strength of an attitude, not just whether an agent has an attitude, that sets this model apart.

Note that there are two timescales here. The first is an inter-agent timescale where the protocol defined above is followed. Within each "tick" of this timescale is an updating of the cognitive network of the hearer according to Eq. (3) which is the second timescale. The timescale will be apparent from the context.

6 Experiments

To explore attitude resiliency and agreement we have implemented the model discussed above using a simple, manually designed cognitive network, and have executed numerical simulations while varying the distribution of attitudes among the agents. We would like to see whether the population can agree on a particular attitude and how long it takes to do so.

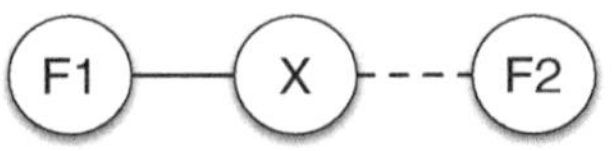

Fig. 3 Cognitive network of the agents in the simulation

6.1 Cognitive Network

Since our objective is to understand the interaction of populations with different attitudes we designed a simple cognitive network that can represent two conflicting attitudes, shown in Fig. 3.

The concept of interest is the central node labeled X. The two nodes $F1$ and $F2$ represent information or factors that can influence the value of X. $F1$ has a positive connection to X so an increase in $F1$ increases X; $F2$ negatively influences X and thus an increase in $F2$ decreases X. This cognitive network represents a situation where individuals are receiving conflicting information; for instance evidence supporting and rejecting the theory of Anthropogenic Global Warming.[2] The weights are 0.05 and −0.05 for the positive and negative links respectively.

$F1$ and $F2$ do not represent specific pieces of information but rather, in general, all positive and negative information. So a particular piece of information (for instance, "the poloar ice caps are melting") would increase the value of $F1$ or $F2$. This abstraction allows us to focus our attention on the impact of information on X rather than on how to interpret information with regards to a concept.

Although simple, this network does exhibit several important characteristics that we want to capture. First, all concepts have continuous valences thus the ability to represent the strength of belief. Second, the network exhibits bidirectionality—the valence of concept X is influenced by the evidence from $F1$ and $F2$; but also belief in the evidence is influenced by belief in X.

There are two organizations of the nodes that satisfy all the constraints. In the first, which we call the *positive state*, $F1$ and concept X have positive valences while $F2$ has a negative valence. The second is called the *negative state* where $F1$ and attitude X are both negative, and $F2$ is positive. The two states reflect polar opposites—in the positive state individuals believe the positive evidence, reject the negative evidence and believe in concept X. The negative state is the opposite situation where individuals accept the negative evidence and thus reject the positive evidence and reject the concept X. For the positive state we used the values $F1 = 1$, $X = 1$, $F2 = -1$ and for the negative state we used the values $F1 = -1$, $X = -1$, $F2 = 1$.

[2]See: http://climatedebatedaily.com/ for a large set of articles both for and against anthropogenic global warming.

6.2 Experimental Setup and Results

The experimental setup was as follows:

- A population of n agents was initialized with f_p fraction of the population assigned the positive state and $f_n = 1 - f_p$ assigned the negative states.
- The protocol described above was run for t =100,000 number of steps.

Note that due to the decay parameter, the positive and negative states are not attractors under repeated application of the update process. Figure 4 shows the trajectory from the positive/negative states $F1/F2$. Each point represents the valences of $F1/F2$ after an application of the update process. Initially, there is a large jump from the positive/negative state. However, we see a gradual reduction in the amount of change. The two endpoints of the trajectory are designated the positive and negative endpoints and are indicated in the figures below.

To simulate the exchange of information that can impact an attitude we restricted the topics to be only $F1$ and $F2$. X was only modified through the hearer update process. The simulations here are with $n = 484$ agents.

We experimented with four social network topologies: a complete graph, a 2-D lattice, a regular graph and a small world network. The regular graph was of degree 10. Small world (SW) networks have a high clustering coefficient (a higher likelihood that a two neighbors of an agent are also neighbors), while also having a low characteristic path length (mean distance between nodes). These types of networks are known to describe a variety of real world networks, such as the power grid of the U.S. [43]. We generated SW networks using the Watts–Strogatz algorithm with $k = 10$ and $p = 0.01$ [43].

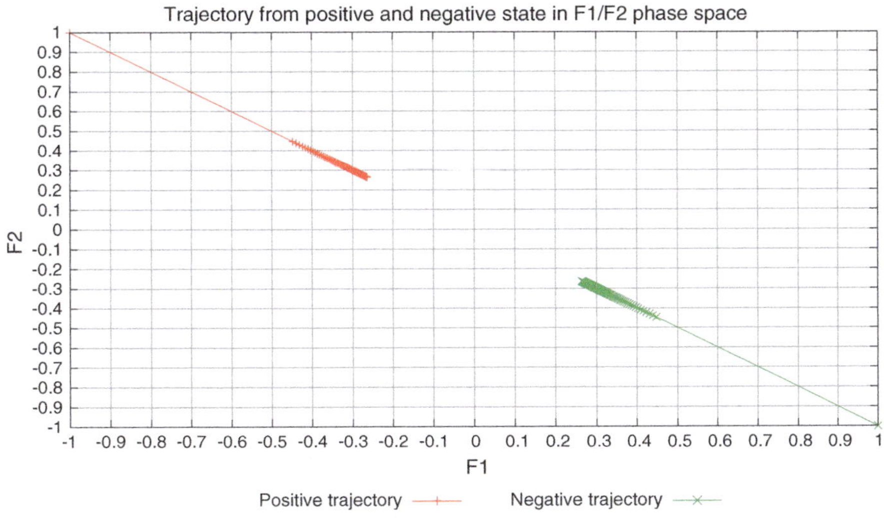

Fig. 4 Trajectory from the positive and negative states over 100 state updates

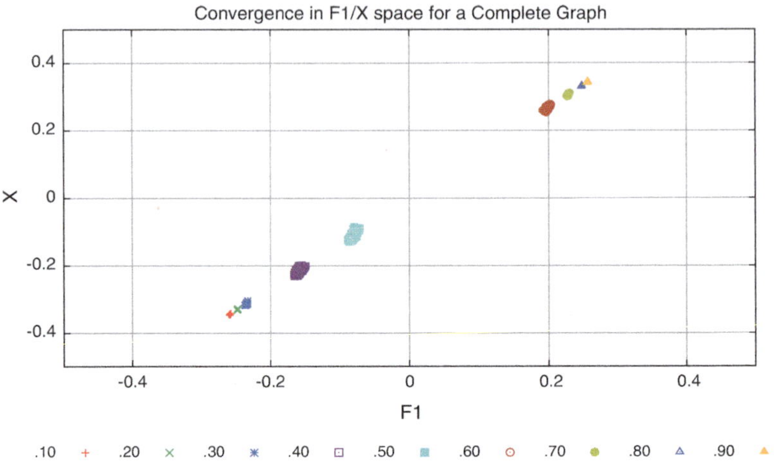

Fig. 5 Location in F1/X space after 99,000 iterations on a complete graph

Unless otherwise stated, results are averaged over 10 runs.

Figure 11 shows the time to reach agreement for the three network topologies topologies as f_p ranges from 0.1 to 0.9. We define agreement as when the mean euclidean distance between the valences of all pairs of agents is less than 2% of the maximum distance. The error bars indicate one standard deviation. Means and standard deviations were calculated based only on the runs that reached agreement. Figure 12 shows the number of runs out of 10 that reached agreement within 100,000 iterations. Notice that for $f_p = 0.5$ the lattice, SW and regular networks did *not* converge within 100,000 iterations and thus no points are provided.

Figure 10 shows the valences of all agents for a range of f_p values in $F1/F2$ space on a complete graph for a single run. The positive and negative end points are indicated as well. Each smudge of color is 484 agents, the color indicates the value of f_p. Figure 5 is the same as Fig. 10 except it shows the valences in $F1/X$ space. Both of the figures combined indicate the valences of the entire population.

Figures 6, 7 and 8 show the end states of all agents for a range of f_p values. End states for $f_p = 0.4, 0.5$ were not shown as system rarely converged in those cases. The presented figures are indicative of the ending state of all the runs.

6.3 Discussion

First, does the system reach agreement, and if so, how long does it take to reach agreement? We can see that in most cases the majority of runs do reach agreement. However, for all non-complete graphs very few runs reached agreement when f_p was close to 0.5. In these cases it seems that the population has split into two groups.

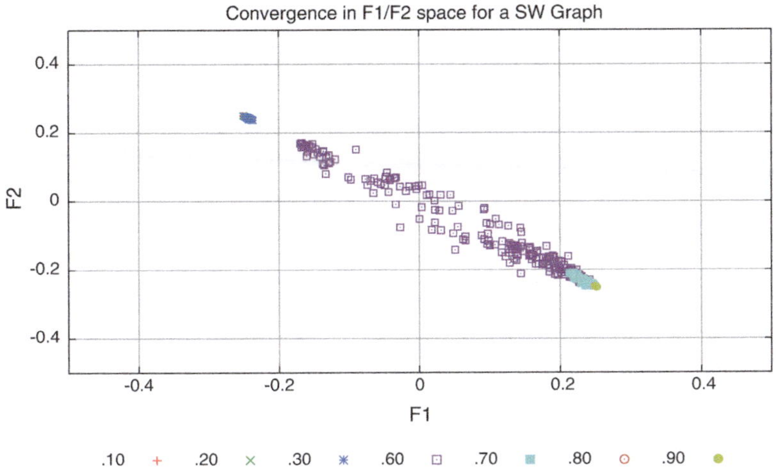

Fig. 6 Location in F1/F2 space after 99,000 iterations on a small world graph

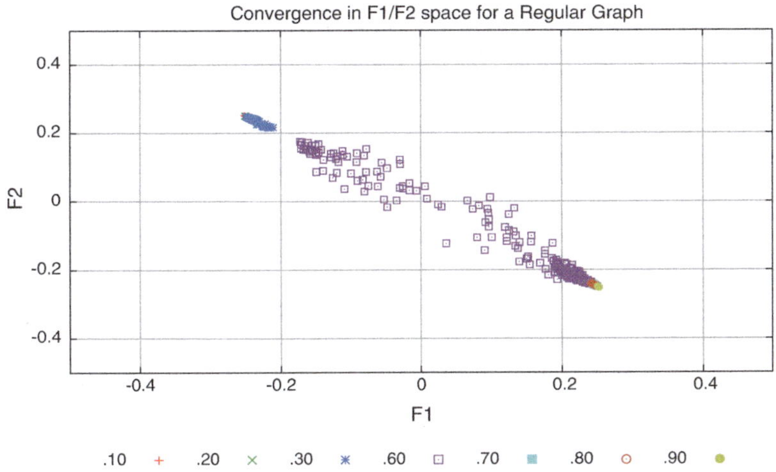

Fig. 7 Location in F1/F2 space after 99,000 iterations on a regular graph

Figure 9 shows the valences of agents in a run that did not result in agreement for the small world network with $f_p = 0.4$. We can see that two groups have formed that do not seem to coming closer.

The lattice, SW and regular graphs take much longer to reach agreement than the complete graph. This has been noted in a variety of models, including the voter [35], and the majority-state update [6] models. However, in our case the small-world networks takes longer than even the lattice and regular networks to reach agreement. This is counter-intuitive, since SW networks have a lower characteristic path length one would expect a decrease in the time to agreement. This is a matter for further

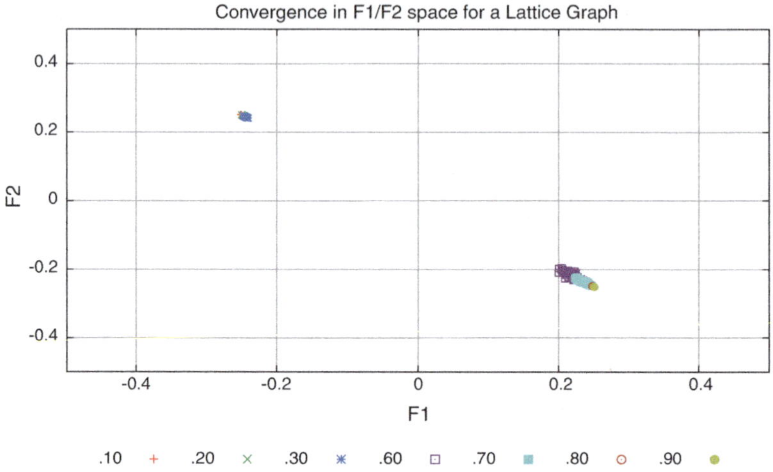

Fig. 8 Location in F1/F2 space after 99,000 iterations on a lattice graph

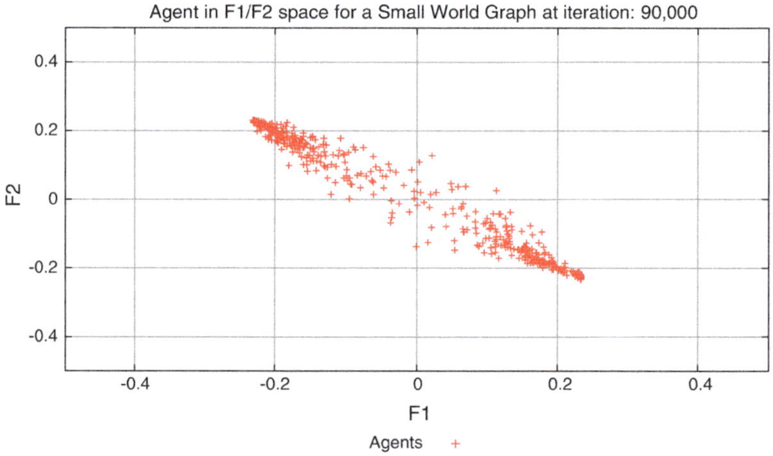

Fig. 9 Agents in $F1/F2$ space for a run that did not result in agreement at iteration 99,000

study. Results for the voter model on SW networks can provide some insight into this behavior as in the thermodynamic limit (where $n \rightarrow \infty$) agreement does not occur [3].

Note the symmetry around $f_p = 0.5$ in terms of time to agreement. This symmetry is also found in voter models for complete graphs [35].

Secondly, what state does the system agree upon? Figures 6, 7, 8 and 10 indicate that the agreed upon state is a function of the topology of the social network and f_p. In the complete graph case we see a linear dependence, as f_p increases the agreed upon state comes closer to the the positive state. However, for non-complete graphs

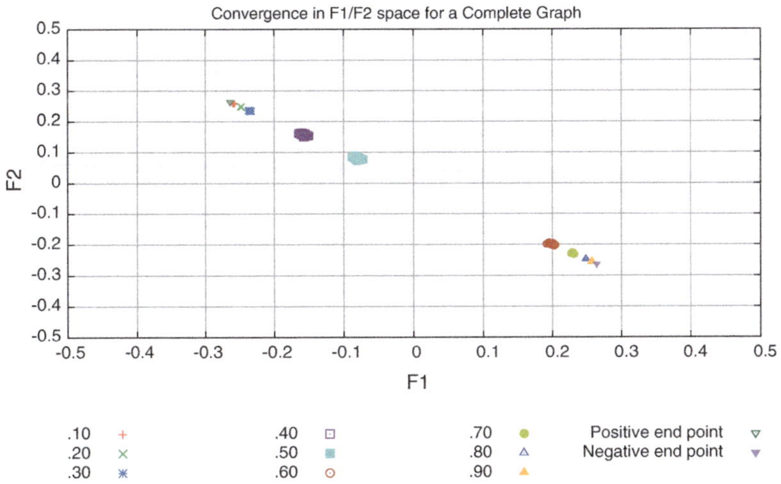

Fig. 10 Location in F1/F2 space after 99,000 iterations on a complete graph. The trajectories from the positive and negative start points are labelled

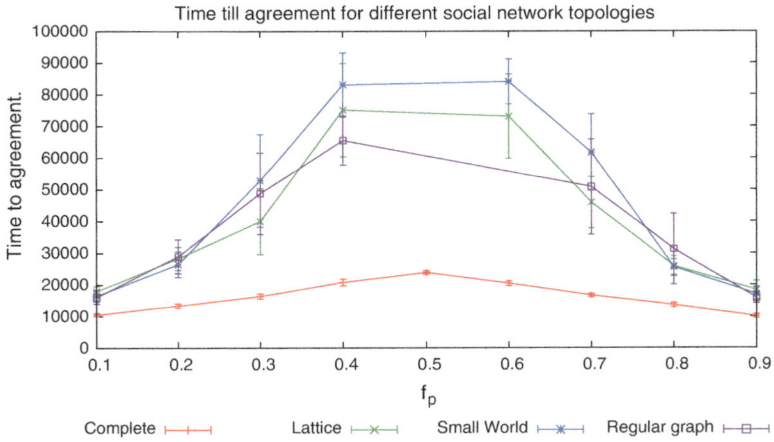

Fig. 11 f_p versus Number of iterations till agreement

the agreed upon states cluster around the positive and negative endpoints. Below $f_p = 0.4$, the systems reach agreement, but very close to the negative state, above $f_p = 0.6$, the system reaches agreement as well, but on the positive state.

These results are especially interesting for the SW networks, as these are many real-world networks. For the simple cognitive network described here, these results indicate that a population will either split into two groups (for $f_p = 0.4, 0.5, 0.6$) or one attitude will overtake the other attitude. Unlike the complete graph case where attitudes change in strength, we see no such variance here (Figs. 11 and 12).

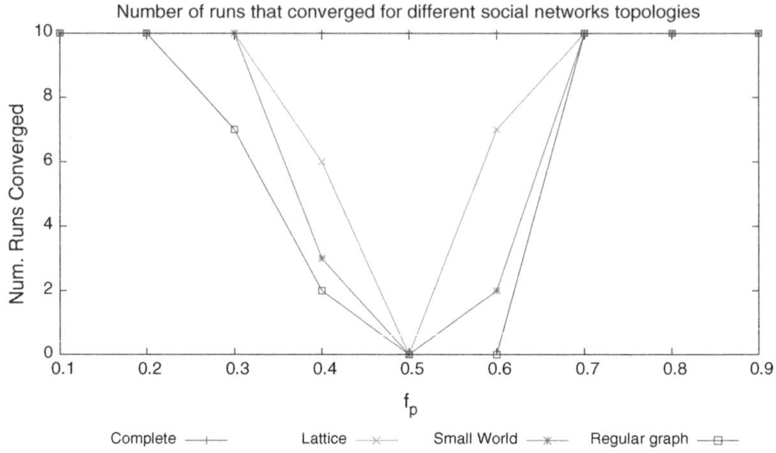

Fig. 12 f_p versus Number of runs that reached agreement

The experiments described here served as a preliminary illustration of the model developed in the prequel. We can see, however, from these simulations several important things:

1. We have shown, through numerical experiments, that our model can capture the process of reaching agreement on a concept, even when there are agents with contradictory attitudes.
2. We have shown that the distribution of attitudes in a population has a large effect on the time to agreement as well as the strength of the attitude. This is interesting as it allows us to study not only whether populations agree, but on what they agree upon.
3. The social network can have a great affect on the final agreement state—for lattices and small-world and regular graphs we see a polarization effect—agents agree on either the positive or negative endstate.

7 Related Work

Many models have been proposed that focus on the social aspect or the cognitive aspect of attitude change; however not many have put together both of these aspects. While a full review of these are out of scope for this work, we refer the interested reader to [2, 12, 17, 21]. Importantly, most models do not account for a complex cognitive aspect for the agent—individuals are often represented as having binary attitudes.

The communication and interaction dynamics of our model are closely related to the *voter model* a well explored model from the physics domain [35]. In this model there are a set of sites each of which is endowed with a variable that can take either state 0 or state 1. At each timestep a random site changes its state to a randomly chosen neighbors. The voter model has been studied on lattices as well as small-world and heterogenous degree networks [35, 38]. Multiple extensions have been developed that incorporate a majority rule (sites change to the state that is in the majority of their neighborhood) [5, 20, 24]. The main issues with voter models are that they are limited to one concept with only two states; this simplicity is, however, a boon for analysis.

Similar to the voter model, at each timestep we choose one agent which communicates with one of its neighbors. However, our agents are more complicated as they utilize a PCS network.

The conformity-consistency model (CCM) is a model that represents both social and cognitive factors [1, 23]. In the CCM, there are N agents, each endowed with a binary vector of size M that represents their cognitive state. At each time step an agent is chosen and it will execute the standard voter model process on of its variables with probability p; with probability $q = 1 - p$ the agent engages in a voter model *with itself*, between the elements of its cognitive state. The voter process within itself can be viewed as reducing dissonance between the elements of its cognitive state—with the extremes of all variables at 0 or all variables at 1 to be the no dissonance situations.

The CCM is very similar to our model, however the representation of the the cognitive state of an agent is significantly limited in only representing binary values. In addition, the model only captures positive relations between concepts; it does not capture negative relations nor does it specify a strength for the relation.

Chapter 10 of [39] describes the "consensus = coherence + communication" (CCC) model. In this model each agent has a PCS network where concepts represent hypotheses and evidence, and links represent explanatory relations. The purpose of the model was to explain the diffusion of scientific theories in a population, so pairs of agents could interact (simulating a complete network) and agents could execute "lectures" for many other agents.

The main difference between our model and the CCC model is in the interaction. In the CCC model agents exchange concepts with others, thus changing the structure of the network. We have focused on a different perspective, given that individuals already know the links between concepts, how do persuasive messages between agents lead to attitude change.

The parallel constraint satisfaction model has been explored in a wide variety of contexts, such as impression formation [16], legal inference [30, 32, 34], and as a model of change in attitude to the persian gulf war [36].

8 Conclusions & Future Work

Attitudes play a significant role in determining how individuals process information and behave. In this paper we have developed a new computational model of population wide attitude change that captures the *social* level: how individuals interact and communicate information, and the *cognitive* level: how attitudes and concept interact with each other. The model captures the cognitive aspect by representing each individual as a parallel constraint satisfaction network.

As a way of understanding the dynamics of this model we have explored a simple attitude change experiment. A population of agents was divided into two groups with opposing attitudes. We wanted to know if the population can reach agreement, how long that would take, and what attitudes would the population reach. We found that: (1) the time to agreement varies with the social network and the initial distribution of attitudes in the population, with the complete graph taking the least amount of time to reach agreement; (2) in complete graphs the final state of the population has a linear dependence with the f_p (the proportion of agents initially with a positive attitude); however for non-complete graphs there was a polarization effect where the population converged on the positive or negative endpoint.

We plan on greatly extending and exploring this model in the future. First, we will try to utilize a more realistic social network. Scale-free networks are a widely known topology that also appear in many physical domains [37].

Developing a more realistic cognitive network is an important part of our future work. We intend to draw from existing literature, for instance from [29] which created a network on breast cancer testing decisions that was developed from survey data.

Acknowledgements Sandia National Laboratories is a multimission laboratory managed and operated by National Technology & Engineering Solutions of Sandia, LLC, a wholly owned subsidiary of Honeywell International Inc., for the U.S. Department of Energys National Nuclear Security Administration under contract DE-NA0003525.

References

1. Bednar, J., Bramson, A., Jones-Rooy, A., Page, S.: Conformity, Consistency, and Cultural Heterogeneity. University of Michigan, Mimeo (2006)
2. Castellano, C., Fortunato, S., Loreto, V.: Statistical physics of social dynamics. Rev. Mod. Phys. **81** (2009)
3. Castellano, C., Vilone, D., Vespignani, A.: Incomplete ordering of the voter model on small-world networks. Eur. Lett. **63**(1), 153–158 (2003)
4. Change, Y.P.O.C.: Climate change in the american mind: Americans' climate change beliefs, attitudes, policy preferences, and actions. Technical report, George Mason University (2009)
5. Chen, P., Redner, S.: Majority rule dynamics in finite dimensions. Phys. Rev. E **71** (2005)
6. Delgado, J.: Emergence of social conventions in complex networks. Artif. Intell. **141**(1–2), 171–185 (2002). http://www.sciencedirect.com/science/article/B6TYF-46RDC1K-5/2/6a3018fe523d5093b5b4ea60f6053ee9
7. Dovidio, J.F., Kawakami, K., Gaertner, S.L.: Implicit and explicit prejudice and interracial interaction. J. Pers. Soc. Psychol. **82**(1), 62–68 (2002)

8. Fazio, R.H., Jackson, J.R., Dunton, B.C., Williams, C.: Variability in automatic activation as an unobtrusive measure of racial attitudes: a bona fide pipeline? J. Pers. Soc. Psychol. **69**, 1013–1027 (1995)
9. Fazio, R.H., Olson, M.A.: Attitudes: foundations, functions, and consequences. In: Hogg, M.A., Cooper, J. (eds.) The Sage Handbook of Social Psychology. Sage, London (2003)
10. Gawronski, B., Bodenhausen, G.: Associative and propositional processes in evaluation: an integrative review of implicit and explicit attitude change. Psychol. Bull. **132**, 692–731 (2006)
11. Gawronski, B., Strack, F.: On the propositional nature of cognitive consistency: dissonance changes explicit but not implicit attitudes. J. Exp. Soc. Psychol. **40**, 535–542 (2004)
12. Goldstone, R.L., Janssen, M.: Computational models of collective behavior. Trends Cogn. Sci. **9**(9) (2005)
13. Hopfield, J.J.: Neural networks and physical systems with emergent collective computational abilities. Proc. Natl. Acad. Sci. USA **79**, 2554–2558 (1982)
14. Jones, C.R.M., Fazio, R.H.: Associative strength and consumer choice behavior. In: Haugtvedt, C.P., Herr, P.M., Kardes, F.R. (eds.) Handbook of Consumer Psychology, pp. 437–459. Psychology Press, New York (2008)
15. Krosnick, J.A.: The role of attitude importance in social evaluation: a study of policy preferences, presidential candidate evaluation, and voting behavior. J. Pers. Soc. Psychol. **55**(2), 196–210 (1988)
16. Kunda, Z., Thagard, P.: Forming impressions from stereotypes, traits, and behaviors: a parallel-constraint-satisfaction theory. Psychol. Rev. **103**(2), 284–308 (1996)
17. Lakkaraju, K.: Agreement, information and time in multiagent systems. Ph.D. thesis, University of Illinois at Urbana-Champaign (2009)
18. Lepper, M.R., Shultz, T.R.: Dissonance. In: Wilson, R.A., Keil, F.C. (eds.) The MIT Encyclopedia of Cognitive Science. MIT Press, Cambridge (1998)
19. Levitan, L.C., Visser, P.S.: The impact of the social context on resistance to persuasion: effortful versus effortless responsees to counter-attitudinal information. J. Exp. Soc. Psychol. **44**, 640–649 (2008)
20. Lima, F., Sousa, A., Sumour, M.: Majority-vote on directed erdos-renyi random graphs. Phys. A **387**, 3503–3510 (2008)
21. Mason, W.A., Conrey, F.R., Smith, E.R.: Situating social influence processes: dynamic, multidirectional flows of influence within social networks. Pers. Soc. Psychol. Rev. **11** (2007)
22. McConnell, A.R., Leibold, J.M.: Relations among the implicit association test, discriminatory behavior, and explicit measures of racial attitudes. J. Exp. Soc. Psychol. **37**, 435–442 (2001)
23. Page, S.E., Sander, L.M., Schneider-Mizell, C.M.: Conformity and dissonance in generalized voter model. J. Stat. Phys. **128**(6), 1279–1287 (2007)
24. Pereira, L.F.C., Moreira, F.G.B.: Majority-vote model on random graphs. Phys. Rev. E **71**(016123) (2005)
25. Read, S.J., Snow, C.J., Simon, D.: Constraint satisfaction processes in social reasoning. In: Proceedings of the 25th Annual Meeting of the Cognitive Science Society (2003)
26. Read, S.J., Vanman, E.J., Miller, L.C.: Connectionism, parallel constraint satisfaction processes, and gestalt principles: (re)introducing cognitive dynamics to social psychology. Pers. Soc. Psychol. Rev. **1**(1), 26–53 (1997)
27. Russo, J.E., Carlson, K.A., Meloy, M.G., Yong, K.: The goal of consistency as a cause of information distortion. J. Exp. Psychol.: Gen. (2008)
28. Rydell, R.J., McConnell, A.R.: Understanding implicit and explicit attitude change: a systems of reasoning analysis. J. Pers. Soc. Psychol. **91**(6), 995–1008 (2006)
29. Shoda, Y., Mischel, W.: Personality as a stable cognitive-affective activation network: characteristic patterns of behavior variation emerge from a stable personality structure. In: Read, S.J., Miller, L.C. (eds.) Connectionist Models of Social Reasoning and Social Behavior. Lawrence Erlbaum, Mahwah (1998)
30. Simon, D.: A third view of the black box: cognitive coherence in legal decision making. Univ. Chic. Law Rev. **71** (2004)

31. Simon, D., Holyoak, K.J.: Structural dynamics of cognition: from consistency theories to constraint satisfaction. Pers. Soc. Psychol. Rev. **6**(6), 283–294 (2002)
32. Simon, D., Krawczyk, D.C., Holyoak, K.J.: Construction of preferences by constraint satisfaction. Psychol. Sci. **15**(5), 331 (2004)
33. Simon, D., Snow, C.J., Read, S.J.: The redux of cognitive consistency theories: evidence judgments by constraint satisfaction. J. Pers. Soc. Psychol. **86**(6), 814–837 (2004)
34. Simon, D., Krawczyk, D.C., Bleicher, A., Holyoak, K.J.: The transience of constructed preferences. J. Behav. Decis. Mak. **21**, 1–14 (2008). https://doi.org/10.1002/bdm.575
35. Sood, V., Antal, T., Redner, S.: Voter models on heterogenous networks. Phys. Rev. E **77** (2008)
36. Spellman, B.A., Ullman, J.B., Holyoak, K.J.: A coherence model of cognitive consistency: dynamics of attitude change during the persian gulf war. J. Soc. Issues **49**(4), 147–165 (1993)
37. Strogatz, S.H.: Exploring complex networks. Nature **410**(6825), 268–276 (2001). http://dx.doi.org/10.1038/35065725
38. Suchecki, K., Eguiluz, V.M., Miguel, M.S.: Voter model dynamics in complex networks: role of dimensionality, disorder and degree distribution (2008). arXiv:Cond-mat/0504482v1
39. Thagard, P.: Coherence in Thought and Action. Life and Mind: Philosophical Issues in Biology and Psychology. MIT Press, Cambridge (2000)
40. Visser, P.S., Clark, L.M.: Attitudes. In: Kuper, A., Kuper, J. (eds.) Social Science Encyclopedia. Routledge, London (2003)
41. Visser, P.S., Cooper, J.: Attitude change. In: Hogg, M., Cooper, J. (eds.) Sage Handbook of Social Psychology. Sage Publications, Los Angeles (2003)
42. Visser, P.S., Mirabile, R.R.: Attitudes in the social contex: the impact of social network composition on individual-level attitude strength. J. Pers. Soc. Psychol. **87**(6), 779–795 (2004)
43. Watts, D.J., Strogatz, S.H.: Collective dynamics of 'small-world' networks. Nature **393**(6684), 440–442 (1998). http://dx.doi.org/10.1038/30918
44. Wilson, T.D., Lindsey, S., Schooler, T.Y.: A model of dual attitudes. Psychol. Rev. **107**(1), 101–126 (2000)

An Application of Agent Based Social Modeling in the DoD

Catherine Zanbaka, Jason HandUber and Desmond Saunders-Newton

Abstract PMRFSim is a proof of concept geospatial social agent-based simulation capable of examining the interactions of more than 60,000 agents over a simulated year within a few minutes. PMRFSim utilizes real world data from federal, state, and local authorities to create the heterogeneous agent environment. All agents are endowed with demographic and geospatial attributes. Agents interact with each other and respond to events. PMRFSim also models adverse agents which have harmful intent and goals to spread negative sentiment and acquire intelligence. PMRFSim allows an analyst to construct various what-if scenarios and generates numerous graphs that characterize the social landscape. This analysis is intended to support public affairs officers in exploring the social landscape under various initial conditions and scenarios.

1 Introduction

A situational awareness capability is the ability of an individual, organization or system to understand and assess, as well as act upon or respond to, its environment within context. For performers in the military, situational awareness has been viewed as a capability that affords operators and analysts to act more effectively in their environment in order to avoid, mitigate or exploit actions taken by others. Long-term situational awareness enables threat anticipation, which allows decision makers to expend fewer resources by addressing a situation earlier.

C. Zanbaka (✉)
Berytus Tech, LLC, 2921 Cravey Drive, Atlanta, GA 30345, USA
e-mail: czanbaka@gmail.com

J. HandUber
BAE Systems, 600 District Ave, Burlington, MA 01803, USA
e-mail: Jason.HandUber@baesystems.com

D. Saunders-Newton
School of Policy, Planning and Development, University of Southern California, Los Angeles, CA, USA
e-mail: desmondsn@gmail.com

T. Carmichael et al. (eds.), *Complex Adaptive Systems*, Understanding Complex Systems, https://doi.org/10.1007/978-3-030-20309-2_3

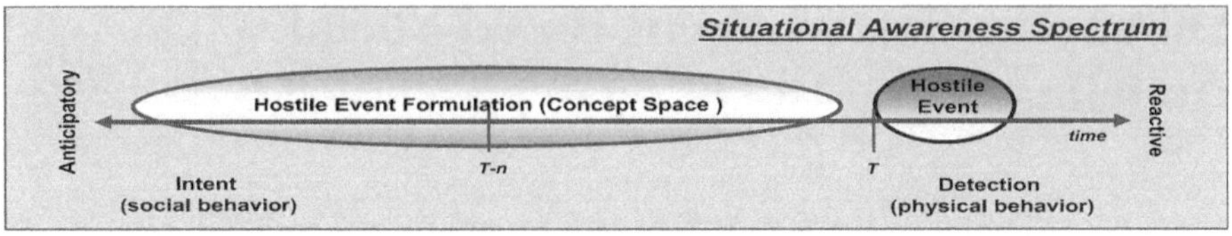

Fig. 1 Situational awareness spectrum

Improvements in situational awareness can occur along a temporal spectrum delineated by the extremes of reaction and anticipation, as depicted in Fig. 1. It is preferred to shape the situation so as to avoid responding to a negative event. Thus, the goal is to detect conditions that facilitate hostile intent thereby reducing, as early as possible, the likelihood of hostile action. This approach encourages an anticipatory rather than a reactive approach to base security. We implemented a computational social model to aid in social situational awareness at the intent formation end of the Situational Awareness Spectrum for the Pacific Missile Range Facility (PMRF), located on the island of Kauai in Hawaii.

Kauai's relatively small population, domestic location, and somewhat isolated environment present an ideal region to conduct social science research. PMRF Simulator (PMRFSim) provides insights into the security policy and its effects on the local population's perception of and interaction with PMRF. PMRFSim can aid in identifying potential 'tipping points' in the population's sentiment, allowing for correction *before* a hostile event takes place.

PMRFSim provides a computational social environment that enables the study of the following aspects of base protection:

- Social responses to user specified events
- Acquisition and dissemination of physical security information
- Spread of positive and negative sentiment towards PMRF.

2 Related Work

PMRFSim builds upon the general agent-based modeling paradigm. Based largely on the capabilities that emerged with the object-oriented programming approach, this paradigm affords researchers from a variety of disciplines with an opportunity to rethink how to instantiate ideas—or theories of note—as algorithms. This enabled social scientists to examine social and behavioral phenomenology at the level of "agency" as opposed to that of "aggregates." Most quantitative methodologies, e.g. multivariate statistics and closed-form analytic forms, focus on summary statistics of large populations which does not allow researchers to easily consider heterogeneous populations characterized by large amounts of interactions between the individual agents and their environment.

Past work of note includes the famous Gedanken experiment performed by Thomas Schelling to explore segregation [11], which is easily represented as a cellular automata or agent-based model in silico, and which demonstrated how individual (local) choices can give rise to interesting global outcomes. This ability to easily reflect multi-level social phenomenon, i.e. micro-level actions emerging as macro-level outcomes, is one of the strengths of the agent-modeling paradigm. These capabilities were further enhanced in the early 1990s, with research in nonlinear dynamics and advances in algorithmic representation, as illustrated by the work of Epstein and Axtell [4].

Since the publication of that book, which illustrates how to reflect the principles of social science theory as agent (object)-level interaction rules and entity attributes, a number of research efforts have been pursued in order to demonstrate the usefulness of this modeling approach to issues of global security and national defense planning. Efforts related to PMRFSim in terms of methodological perspectives and study focus include the civil violence model developed by Epstein [5], the Defense Threat Reduction Agency funded work in threat anticipation by Los Alamos researcher Ed MacKerrow [7], and Cultural Dissemination/Diffusion models developed by Robert Axelrod of the University of Michigan [1].

In focusing on situational awareness and intent formation, PMRFSim has taken advantage of advances in geospatial representation, social interaction theory, and opinion dynamics [9, 12], as well as influence/memetic propagation [10]. PMRFSim has also benefited from advances in how to "use" advanced analytic technologies more credibly as discussed by Frank and Saunders-Newton [6].

3 Input Data

PMRFSim uses real-world data inputs to define the Kauaian population. These real world inputs include geographic and demographic information from the state of Hawaii and the U.S. Census Bureau, the PMRF security posture, PMRF physical security integrity data, and a region-specific sentiment analysis performed by a local sociologist. PMRFSim uses this real-world data to create and model one agent per person on Kauai, yielding a population of approximately 60,000 agents.

3.1 Geographic Data

PMRFSim uses a combination of editable shapefiles and comma-separated value (CSV) files for geographic data. The files include: a boundary file of the facility and Kauai, a file that contains the local urban areas, guarded areas, open access areas, restricted areas, and a file that contains information that pertains to PMRF that can be used as physical security information (PSI).

3.2 Other Input Data Files

3.2.1 Occupations File

The occupation file specifies occupations by industry and percentage of working population in that industry. All individuals are assigned an occupation, e.g. students, infants, unemployed. The occupations and percentages used for Kauai are listed in Table 1.

3.2.2 Work-Home File

The work-home file is a CSV file that details where people live and work on Kauai. This file is used to assign a home and work location for each agent created in the simulation. The data in this file was retrieved from the Longitudinal Employer-Household Dynamics database which is maintained by the US Census Bureau. Figure 2 shows a connection between each agent's work and home locations on Kauai.

Table 1 Occupations and percentages for Kauai

Industry	Percentages (%)
Agricultural / Forestry / Fishing	0.48
Construction	2.80
Manufacturing	0.54
Transportation / Communications	2.65
Wholesale / Trade	1.22
Retail / Trade	14.97
Finance / Insurance / Realestate	4.97
Services	24.25
Public administration	3.63
Preschool	1.66
Kindergarten	1.41
Elementary school	14.26
High school	7.69
College	3.31
Unemployed	2.85
Retired	8.09
Infants	5.20

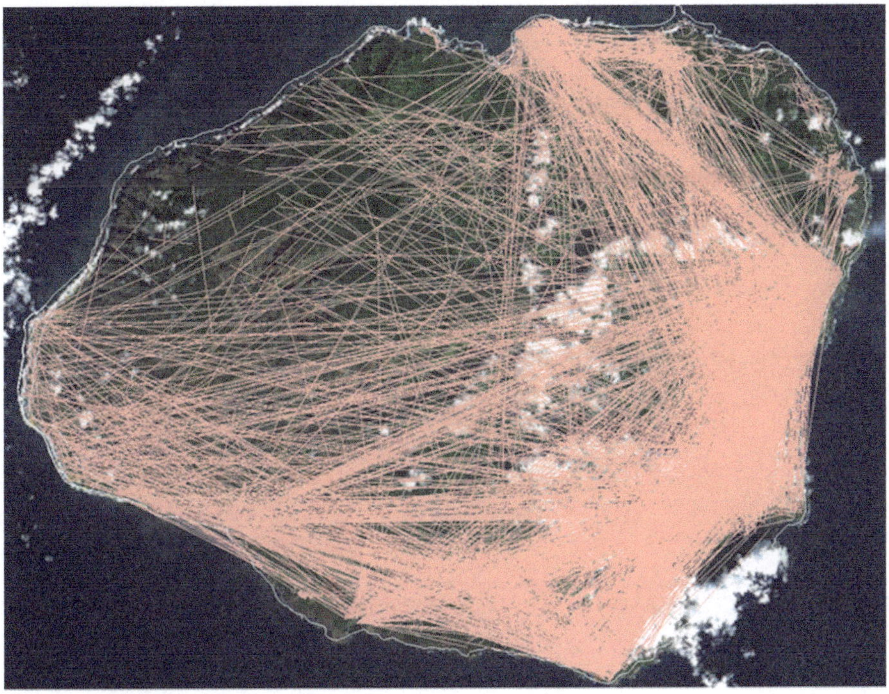

Fig. 2 Work home mappings overlaid on Kauai

3.2.3 Favorite Places File

The favorite places shapefile is used to relocate agents to urban or recreational areas during their free time. Agents will spend most of their time at home and at work, but after work hours and during the weekends, agents are free to go to their favorite places. Each agent is assigned an average of 15 favorite places with a minimum of 5 favorite places, a maximum of 30, and a standard deviation of 3.75. Favorite places also have associated probabilities of visit. The probability of visiting a favorite location is proportional to the distance from that favorite place to either the agent's work or home location (whichever is closer). Figure 3 shows the favorite places shapefile shown within the Kauai boundry.

3.2.4 Census Data File

Agents in the model are created based on a census shapefile which is specified as a map layer. The shapefile was retrieved from the US Census Bureau and includes a region-by-region breakdown of the population by race, age, gender, household size,

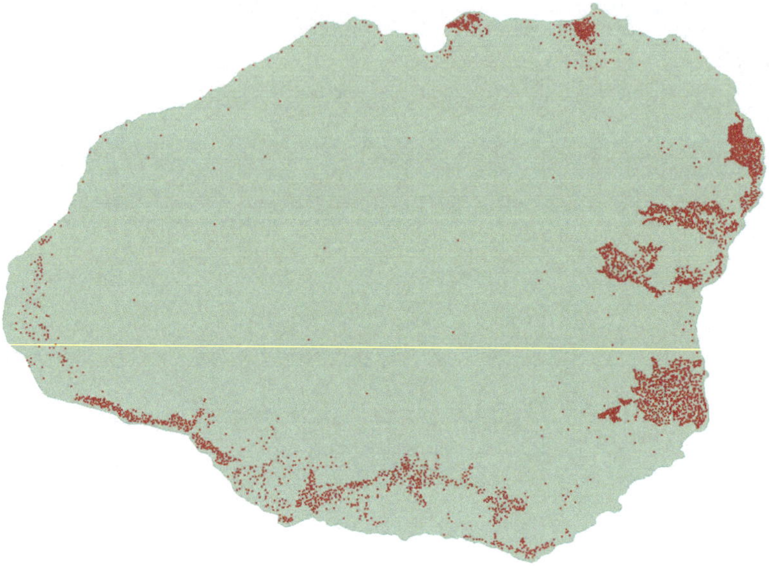

Fig. 3 Favorite places shapefile shown within Kauai boundry

and average persons per household. The census shapefile used in PMRFSim has a total of 58,395 agent definitions.

3.2.5 Topic Interests File

Sentiment analysis utilizes the study of linguistics and language detection processes in order to draw conclusions about a writer or speaker's attitude towards their addressed topic. Sentiment analysis uncovered the topics listed below as being most relevant to local sentiment towards PMRF.

- *Community Support/Outreach*: Personnel from PMRF having an impact on the community through educational guidance, natural disaster relief/assistance, or financial assistance.
- *Economic Incentives/Stimulation*: PMRF is contributing federal monies to the local economy or is attracting tourists to stimulate the local economy.
- *National Security/Defense*: Tests conducted at PMRF help to protect people from external threats or other dangerous actions against citizens on the local and national levels.
- *Operation Test Success*: Scheduled test was successfully performed.
- *Operation Test Failure*: A scheduled test was unsuccessfully performed.
- *Environmental/Safety Concerns*: Identifies significant environmental and safety benefits or risks of PMRF.

- *Technological Advancements*: Discusses significant strides that PMRF is making to progress technology.
- *Public Access to Beaches*: Identifies issues that either hinder or encourage beach access to the public.
- *Public Advocacy/Participation*: Describes the effectiveness of public's voice in the operations and decision making of PMRF.
- *Land Consumption/Control*: Identifies that PMRF is controlling state land which could potentially be alternatively used.
- *Transparency of Decisions*: Discusses the military's operations and decision making with regard to whether PMRF is declassified/open to the public.
- *Militarization of Hawai'i*: Discusses the military's occupation and presence within the state of Hawai'i, whether positive or deleterious.

These topics define different issues of interest for various agent types in the population. For example, someone may be very interested in land use and environmental concerns, but not interested in the other topics. For this particular agent, news about economic stimulation may not affect its sentiment towards the base, but a takeover of public farms by the base would be very important to this agent. These topic interest specifications are used in modeling agent reactions to events.

4 Model Dynamics

PMRFSim simulates each agent and its interactions with other agents as well as reactions to events. Each agent in the simulation goes to work, home, they sleep, they interact with each other, and they respond to events. PMRFSim events are defined relative to topics provided by a regional analyst. PMRFSim uses social interaction theory and social opinion dynamics to model the individual interactions and reactions to events in the heterogeneous Kauaian population.

Creating the population requires demographic data, occupation data, sentiment analysis, and population composition data. The sentiment analysis then seeds the sentiment model with topic interests and initial sentiment values. The sentiment model is influenced by interacting with other agents, reacting to events, and interactions with PMRF security. In turn, interacting with other agents is dependent on demographic, occupational, spatial proximity, and topic interest data, as well as current sentiment.

Interaction with other agents plays into sentiment and physical security information (PSI) exchange. For adverse agents, these exchanges then feed into the relocation logic thereby influencing the probability of interaction. For normal agents, relocation is a function of the current simulation time, work and home locations, and their favorite places.

4.1 Sentiment

Sentiment is a measure of the attitude of an individual towards an object (i.e. PMRF). This attitude includes two components, (1) the cognitive context, i.e. the topic through which the individual is interested in PMRF, and (2) the affective orientation, i.e. the positive or negative indicator of feeling towards PMRF within the context of the topic. Every agent has a sentiment value between -1.0 (very unfavorable) and +1 (very favorable), with 0 being neutral. For example, an environmentalist agent may be defined as having a relatively strong interest in the environmental topic and thus this agent's sentiment towards PMRF will be largely driven by the agent's perception of PMRF's environmental policy. Agent topic-interests are further discussed below in the Agent Generation section.

The primary focus of PMRFSim is to study and explore Kauaian sentiment towards PMRF. Each modeled Kauaian, or agent, can change its sentiment towards PMRF by (1) interacting with other agents that have different opinions about PMRF, (2) by responding to local events, or (3) by directly interacting with PMRF security. PMRFSim enables an analyst to define different event scenarios and observe the resulting sentiment change in the Kauaian population (e.g. new environmental policy event).

4.2 Agent Generation

PMRFSim models two types of agents: normal and adverse agents. Adverse agents are distinct in that they spend all of their free time attempting to either lower sentiment or acquire PSI.

Each agent is defined by seven scalar values (gender, age, race, occupation, work, home, initial sentiment) and three vectors, one to define the topic weights, one to define the favorite places, and one to define the initial PSI. The topic weights dictate how much an agent cares about each topic. The favorite places vector defines where normal agents go on their free time. The initial PSI vector is typically empty; however, agents that work or live on PMRF have a 30% probability of receiving 'invisible' PSI at the beginning of the simulation. The following data sources are used in agent generation:

- Gender, age, and race by geographic region from the Census Bureau
- Work and home locations from the Census Bureau and State of Hawaii
- Occupation data from the Labor Department
- Initial sentiment from Dr. Sun-Ki Chai's (Regional Sociologist) sentiment analysis
- Topic weights from Dr. Sun-Ki Chai's sentiment analysis
- Favorite places defined in a shapefile
- Initial PSI, distributed randomly to PMRF personnel.

4.3 Agent Interactions

When agents in the PMRFSim model interact, they exchange physical security information (PSI) and influence each other's sentiments about PMRF.

4.3.1 Bayesian Network

In PMRFSim, adverse agents are endowed with a decision making capability by utilizing Bayesian networks for uncertainty. A Bayesian network (or a belief network) is a probabilistic graphical model that represents a set of variables and their probabilistic dependencies [2]. A Bayesian network can capture cause and effect relationships, such as "Cloud" yields a likelihood for "Rain". Bayesian belief networks can be used to make inferences about current situations and individual behaviors. This type of reasoning allows the agents to determine an action based on current situational state.

In PMRFSim, the Adverse Agent's Bayesian network is geographically neutral and allows all Bayesian deductions to persist independent of the location of the agent. That is, regional data is recorded in each adverse agent's grid and used by the Bayesian network, but the network itself doesn't record region-specific information (it determines the attractiveness of a region given region-specific information). This dramatically increases the computational efficiency of the Bayesian network. The Bayesian network is used to decide where an adverse agent should physically relocate during that agent's free-time.

There are two types of Adverse Agents in the model, those that primarily seek to lower sentiment, and those that primarily seek to acquire PSI. The decision is based on how much of an effect the agent had on either sentiment at that location or how much PSI was gained at that location. Therefore, the adverse agent keeps track of the following information:

1. *Sentiment Effect at this location*: The amount of agent sentiment in this particular grid cell was affected by the adverse agent.
2. *Knowledge Acquired at this location*: The amount of new PSI the agent has been able to collect at this particular location.

This information is then used when the agent is choosing a location to move to in the simulation. A Bayesian Inference File is required to generate the Bayesian network. In PMRFSim, two Bayesian Inference Files have been created for the two types of Adverse Agents:

1. Maximize Knowledge Acquisition
2. Maximize Sentiment Affect.

4.3.2 Probability of Interaction

Agents in PMRFSim choose their interaction partner using a probability of interaction equation (Eq. 1) based on the work of Moon and Carley [9]. This equation takes into account both physical distance and relative similarity between two agents. Using similar attributes for deciding on an interaction partner follows the principle of homophily [8] and similarity attraction theory [3] from social science.

To account for spatial proximity, PMRFSim considers a maximum of 15 randomly drawn agents from within the same grid cell. The space is broken down into grid cells (1 km × 1 km by default, resulting in ~2184 cells) to provide significant computational time savings without sacrificing the plausibility of the simulation.

Once the agent has a list of potential interaction partners, the probability of interaction for each is computed. The interaction partner with the highest probability is chosen for interaction.

The Probability of Interaction equation used in PMRFSim (Eq. 1) takes into account demographic similarity and sentiment (DSS) as well as shared topic interests (STI) between 2 agents. Both DSS and STI are shown in Eq. 2. The DSS between two agents is a function of race, gender, occupation, age, and current sentiment. STI is the percentage of topic interests that both agents share.

$$P_i = \frac{1}{2} DSS + \frac{1}{2} STI \tag{1}$$

$$\begin{aligned} DSS &= \frac{\Delta Race + \Delta Gender + \Delta Occupation + \Delta Age + \Delta Sentiment}{5} \\ STI &= \frac{\#\text{Shared Topic Interests}}{\#\text{Topics}} \end{aligned} \tag{2}$$

4.3.3 Agent to Agent Interactions

As agents interact in the simulation they influence each other's opinions about PMRF. This is done using a variation of the continuous social opinion dynamics algorithm developed by Weisbuch [12]. The Weisbuch equation (Eq. 3) is used to calculate the sentiment of two agents after they interact ($a_1{}'$ and $a_2{}'$ are the respective resulting sentiments, a_1 and a_2 are the initial sentiments and μ is a constant convergence parameter). The convergence constant was tuned based off of guidance from the Weisbuch paper and experimentation.

$$\begin{aligned} a_1{}' &= a_1 + \mu \times (a_2 - a_1) \\ a_2{}' &= a_2 + \mu \times (a_1 - a_2) \end{aligned} \tag{3}$$

Our variation of the above equation, Eq. 4, is asymmetric in the sense that the convergence rate is not necessarily equal for both agents. Rather, the convergence rate is related to the strength of the agent's opinion. An agent that feels strongly about a topic is harder to convince than one that is undecided. Rather than using a fixed value for convergence, we compute it for each agent. Below is the formula, where a_1 and a_2 are the opinions of two agents, and μ_1 and μ_2 are their respective calculated convergence parameters, CR_1 and CR_2 are each agent's immutable convergence parameter, and $a_1{}'$ and $a_2{}'$ are the agent's opinions after the interaction.

$$\begin{aligned} \mu_1 &= 1 - |a_1| \times CR_1 \\ \mu_2 &= 1 - |a_2| \times CR_2 \\ a_1{}' &= a_1 + \mu_1 \times (a_2 - a_1) \\ a_2{}' &= a_2 + \mu_2 \times (a_1 - a_2) \end{aligned} \tag{4}$$

4.3.4 Agent to Event Interactions

PMRFSim provides a topic-centric event mechanism. Using this mechanism the user can define an event by defining an associated anticipated affect that event will have on each agent, by topic. Table 2 shows a sample event that has a very positive economic effect, slightly negative environmental consequences, positive national defense implications, and slightly positive effect on technology. Many topics are related, for example economic events are usually accompanied by an impact on the environment or technology.

PMRFSim uses a modified version of Eq. 4 for opinion change in response to an event. A higher convergence rate is used for opinion change in response to an event than that used for agent to agent opinion change.

An agent's sentiment in response to an event is only affected if that event touches on topics that the agent is interested in. The total change in sentiment is a sum of changes in response to each topic, as illustrated in Eq. 5.

$$S' = S + \sum \mu \times A_T \times (E_T - A_S) \tag{5}$$

where A_T is the agent topic weight, A_S is the agent sentiment, E_T is the event topic weight, and μ is the agent to event convergence rate.

To illustrate the agent to event social opinion equation, the example in Eq. 6 shows Agent A, defined in Fig. 4, receiving the economic event defined in Table 2. The agent's total sentiment change (S') is the sum of his response to the economic, environmental, national defense, and technology aspects of that event. This calculation shows that user defined events have a strong impact on sentiment when both: (1) the agent is very interested in one of the event topics and, (2) the sentiment of

Table 2 Sample event definition

Economy	Environment	Nat'l Defense	Technology
0.8	-0.1	0.4	0.1

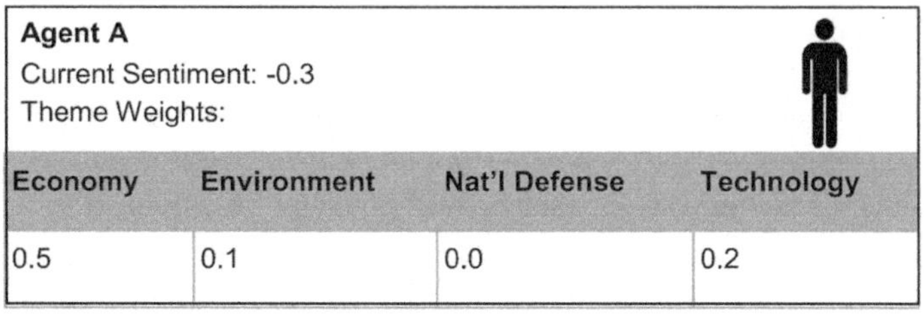

Fig. 4 Sample agent sentiment and topic interests

the agent and the anticipated sentiment of the topic-of-interest significantly differ. In essence, an event will only change an agent's sentiment if at least one of its topics expresses a view different from the agent's own view.

$$
\begin{aligned}
&S_{economy} = 0.05 \times 0.5 \times (0.8 - -0.3) = 0.0275\\
&S_{environment} = 0.05 \times 0.1 \times (-0.1 - -0.3) = 0.001\\
&S_{nat'l\ defense} = 0.05 \times 0.0 \times (0.4 - -0.3) = 0\\
&S_{tech} = 0.05 \times 0.2 \times (0.1 - -0.3) = 0.004\\
&S' = -0.3 + (0.0275 + 0.001 + 0 + -0.004)\\
&S' = -0.3 + 0.245\\
&S' = -0.2755
\end{aligned}
\tag{6}
$$

4.3.5 Agent to Base Security Interactions

Every time a non-PMRF agent enters PMRF, the simulation models various possible interactions between the visiting-agent and PMRF security. The assumed sentiment shift resulting from each interaction is shown in Table 3 and are used by Eq. 7

Table 3 Sentiment shifts resulting from base interactions

Interaction	Description	Sentiment Shift
No Contact	Not spotted by security	+0.1
Inspected on Entry	Inspected upon entering PMRF	-0.2
Rejected On Entry Normal	Rejected entry at main gate due to visitor hours or maximum capacity reached	-0.2
Spotted	Spotted by security but no conversation	-0.05
Questioned	Talked to by security, typically a "How are you?" or "Can I help you find something?"	-0.1
Interrogated	Caught by security in a restricted area 4 times or less	-0.4
Banned	Caught by security a 5th time in a restricted area	-0.9
Rejected Entry Banned	Previously banned agent was rejected upon trying to enter PMRF via the Main Gate	-0.8
Jailed	Caught in a restricted area after being banned	-1.0

(a modification of the Asymmetric Weisbuch equation) to calculate the change in sentiment an agent has from a particular interaction with PMRF security. The base interaction social opinion equation (7), calculates S', the agent's new sentiment. S is the agent's starting sentiment, CR is the agent to base convergence rate, and D is the sentiment shift.

$$S' = S + (1.0 - |S|) \times CR \times \Delta \tag{7}$$

4.3.6 Physical Security Information

Physical Security Information (PSI) has a location, a value, a decay rate, and is either visible or invisible. All of these values are specified by the PSI shapefile.

The value of any particular piece of PSI is dependent on how long ago the PSI was collected and the PSI's decay rate. For instance, the location of a moving vehicle patrol may be important, but only for a short period of time. Adverse agents work on a simple Bayesian network using the raw PSI value rather than the aged PSI value. This prevents PSI gathering adverse agents from continually re-visiting the same areas to collect an updated version of the same PSI.

PSI may be visible or invisible. Invisible PSI is distributed to agents that either live or work on PMRF. Visible PSI can be collected by any agent that is within the same base grid cell as the PSI. There is a 50% probability that adverse agents will collect PSI if they are in the same cell and a 10% probability that normal agents will collect PSI if they are in the same cell.

Once all the PSI has been defined it can be divulged to an adverse agent. For example, if a normal agent collects some PSI in January and February and meets an adverse agent in March and the two agents interact, then special logic is invoked to determine what PSI, if any, is divulged to the adverse agent. The first portion of this logic determines which pieces of PSI may be divulged (Fig. 5). The second portion of logic loops over this list and, with a probability equal to the relative similarity between the two agents, divulges the instance piece of PSI to the adverse agent. Note that only adverse agents acquire PSI through interactions.

4.4 Agent Relocation

All agents follow a *sleep* → *work* → *free-time* → *sleep* weekday cycle. On weekends all agents sleep and have free-time. Tables 4 and 5 detail where and when agents go to various locations. For example, the first row in both tables shows that between 2000 and 0800 (next day) all agents should go home (usually arriving before 0100) and schedule themselves to arrive at work sometime between 0800 and 0930. It is in selecting where to go next that adverse and normal agents are distinct, as highlighted by the 'Special Location' reference in the last row of Table 5.

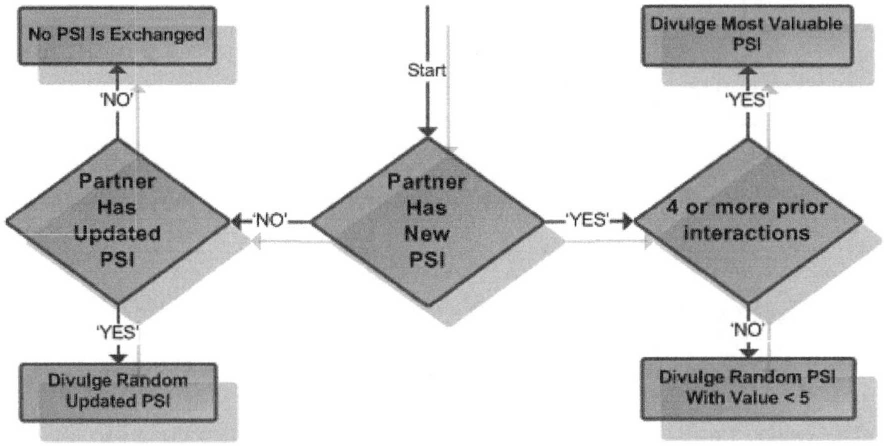

Fig. 5 Generating the list of potential PSI to divulge to an adverse agent

Table 4 Agent movement times

Current Time	Next Movement Time	Description
2000-0800	Positive Normal Distribution, Mean: 0800, Sigma: 30, Max: 930	All agents arrive at work between 0800 and 930
0800-1700 weekdays	Normal Distribution, Mean: 1700, Sigma: 60, Min: 1500, Max: 1900	All agents leave work between 3pm and 7pm
Weekend or after 5pm	Normal Distribution, Mean: *Mean Hours Between Relocations*, Sigma: 1 Hour, Max: 7 Hours	All agents move around, by default, every 4 hours during the weekend or after work

Table 5 Agent movement destinations

Current Time	Next Movement Destination	Description
2000-0800	Home	All agents are at their home location
0800-1700 weekdays	95% Probability of Being At Work, Otherwise: Favorite Location	Agents usually go to work during work hours; otherwise they go to one of their favorite places
Weekend or after 5pm	Favorite Location for normal agents, '*Special Location*' for adverse agents	Normal agents spend their weekends at their favorite places and adverse agents go to places in which they have had success at decreasing sentiment or gathering PSI

4.4.1 Adverse Agent Movement

Adverse agents maintain a history of their performance at previous locations by recording their effectiveness at lowering sentiment and acquiring PSI. This record is not updated as a running mean of how effective the agent was at either goal, but as a window of how effective the agent was the last 5 times the agent visited each cell. Using the window strategy prevents adverse agents from continually going to a single location that provides relatively low sentiment decrease and PSI gain. Instead, adverse agents will explore other places once a location dries up. To select which location to visit, adverse agents utilize a trivial Bayesian network. It is the weights in the Bayesian network that differentiate the two subtypes of adverse agents. Adverse agents use this logic to move to a 'Special Location'.

At the beginning of the simulation, when adverse agents have no record of past performance, they follow the same logic as normal agents and go to one of their favorite places.

5 Analysis

PMRFSim analysis was carried out by defining scenarios of interest and examining the results. The following sections summarize a sample scenario and provide an analysis and interpretation of the results.

5.1 Scenario

The following hypothetical scenario was created by the BAE team for the final demonstration to selected ONR and PMRF staff members in December 2008.

The hypothetical scenario is composed of three key events outlined in Table 6. The scenario uses default values for the social science theory parameters, the Kauaian population, and various simulation parameters.

6 Results

This section provides an interpretation of the results from running the hypothetical scenario outlined in Table 6.

Using multiple output visualizations, PMRFSim provides insights into the social effects of analyst defined events and other PMRFSim inputs. These visualizations illustrate sentiment information over time and space as well as the spread of sensitive information amongst the Kauaian population. Figure 6 shows PMRFSim as it is running, with all graphs and displays enabled.

Table 6 Hypothetical scenario definition

Event	Start Date	End Date	Description	Sentiment Effect
Land Preservation	15 Jan 2008	15 April 2008	PMRF aids in preventing development of lands near northern PMRF	+Environmental +National Defense +Land Control +Public Access
Test Failure	15 May 2008	15 July 2008	A failed missile test results in loss of equipment and injuries	- Test Failure
Economic Growth	15 Sept 2008	31 Dec 2008	PMRF receives a large funding and responsibility increase, bringing economic prosperity to Western Kauai	+ Economic

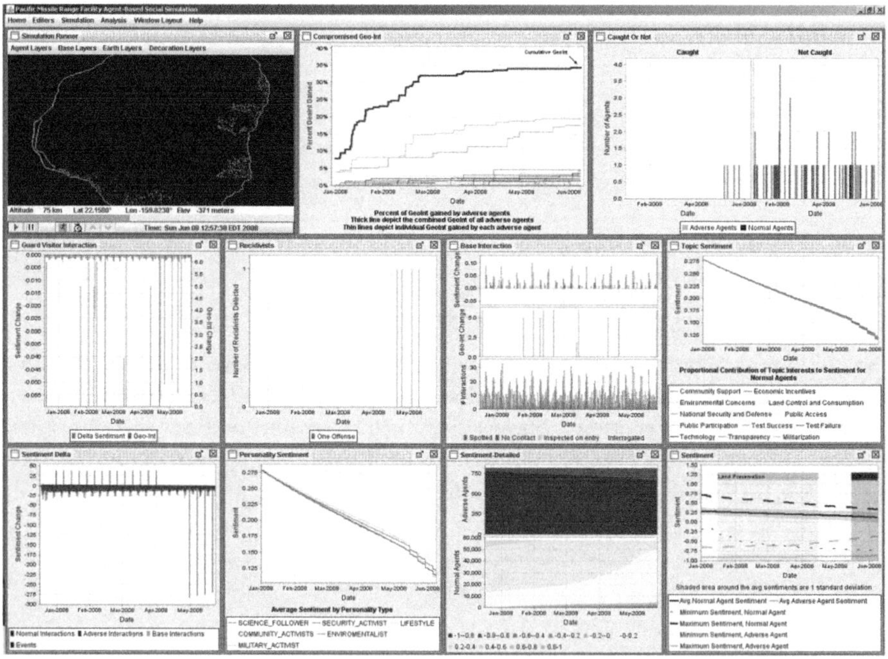

Fig. 6 PMRFSim runtime screenshot

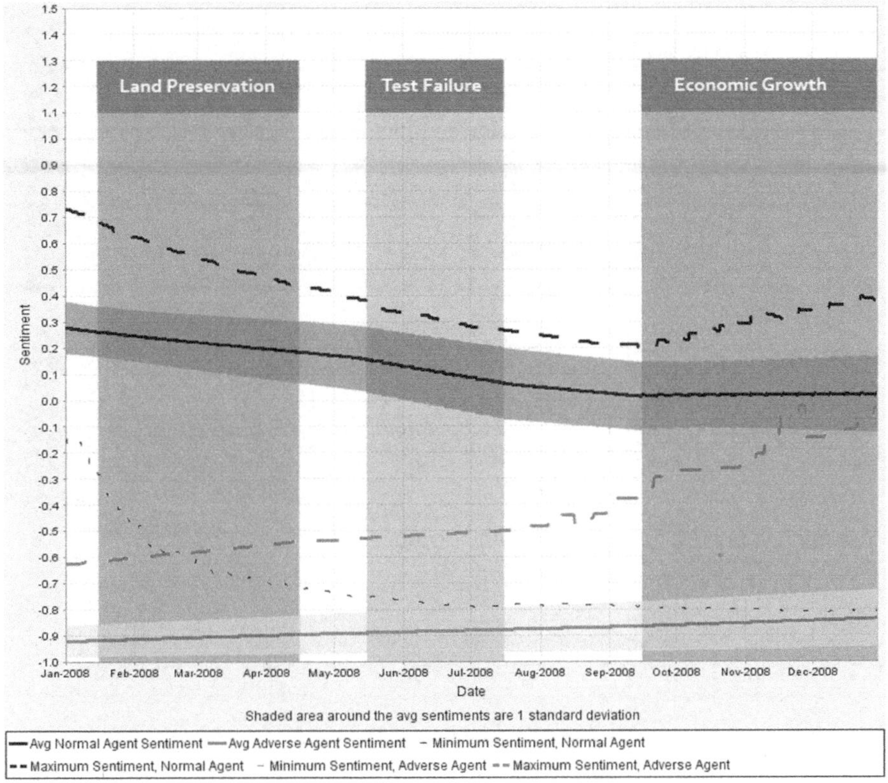

Fig. 7 Minimum, mean, standard deviation, and maximum sentiment by agent type

There are many ways to examine the modeled Kauaian population's sentiment. The next few charts examine Kauaians' sentiment through various visualizations. Figure 7 shows a sentiment graph which shows the highest and lowest sentiment, the average sentiment, and one standard deviation of sentiment for both normal and adverse agents. This graph also overlays the events that were defined during the simulation. Figure 7 shows that, on average, sentiment is only notably changed by powerful events, such as the test failure (2nd event) and economic growth (3rd event) events.

Figure 8 is a detailed sentiment distribution graph and shows how many normal and adverse agents fall into various sentiment bins. From this plot, it is easier to examine the portions of the population that are shifting in sentiment. This figure shows an interesting portion of the population that has been disenfranchised by the simulated events. This disenfranchised sub-population dropped in sentiment after the Test Failure event and their sentiment did not appreciably increase after the positive Economic Growth event. To return the social landscape to its initial state, the issues of this disenfranchised portion of the population must be addressed. The next figure

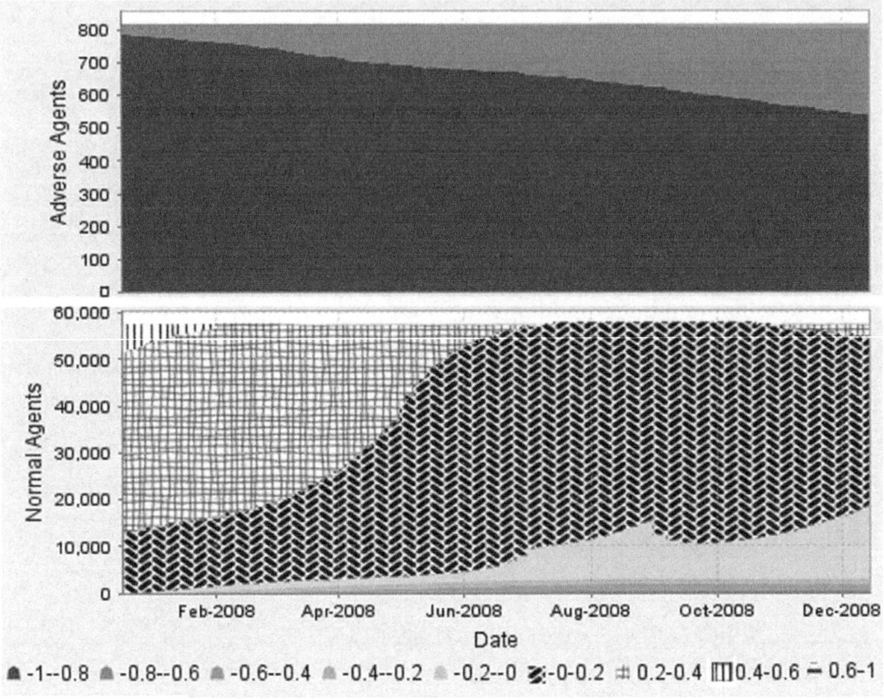

Fig. 8 Adverse and normal agent sentiment binning

will facilitate sentiment examination by identified personality type and will provide further guidance on which personality types may have been disenfranchised thereby providing potential remedies.

Figure 9 is a personality sentiment graph that shows the average sentiment for all agents broken up by personality type. We can see that the military activists are the most unsatisfied after the scenario. Military activists as defined by the sentiment analysis are not concerned at all with the economy, but rather they are most concerned with the topics of the militarization of Hawaii and transparency of the government. It is likely that this personality type makes up a substantial portion of the previously discussed disenfranchised sub-population. Defining an event that improves on militarization and transparency may help the analyst to gain back the support of this portion of the population.

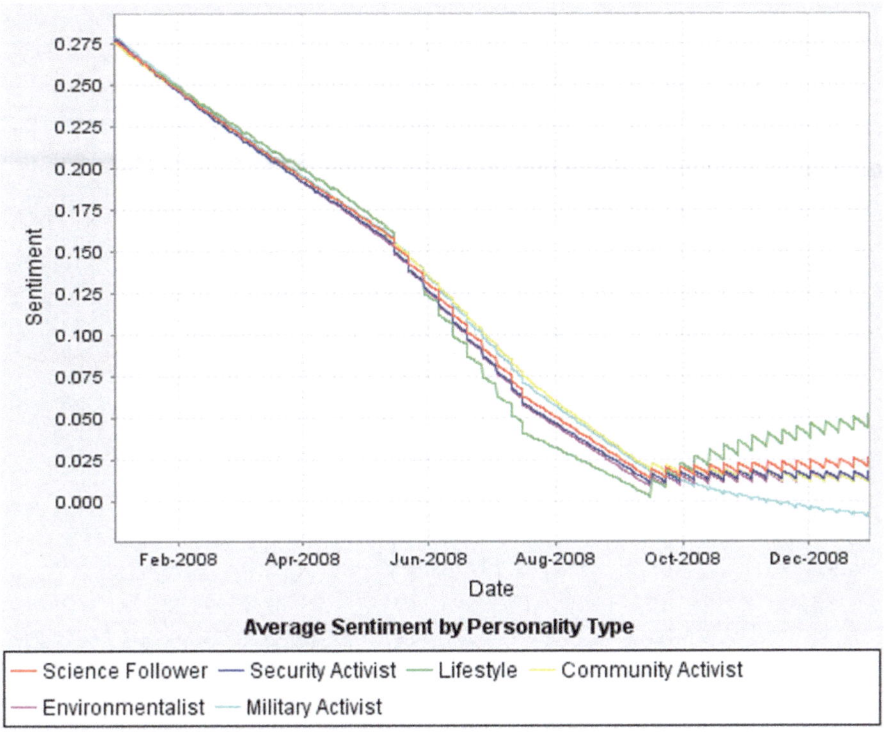

Fig. 9 Average sentiment by personality type

Figure 10 illustrates change in sentiment due to agent interactions with normal agents, adverse agents, base interactions, and responses to events. We can see that the majority of change in sentiment is caused by events, although events do not happen as frequently as interactions. In the simulation event effects are computed on a weekly basis, giving rise to the spaces between the event sentiment spikes seen in Fig. 10. These charts are useful in comparing real-world sentiment shift data against simulation data and assumptions.

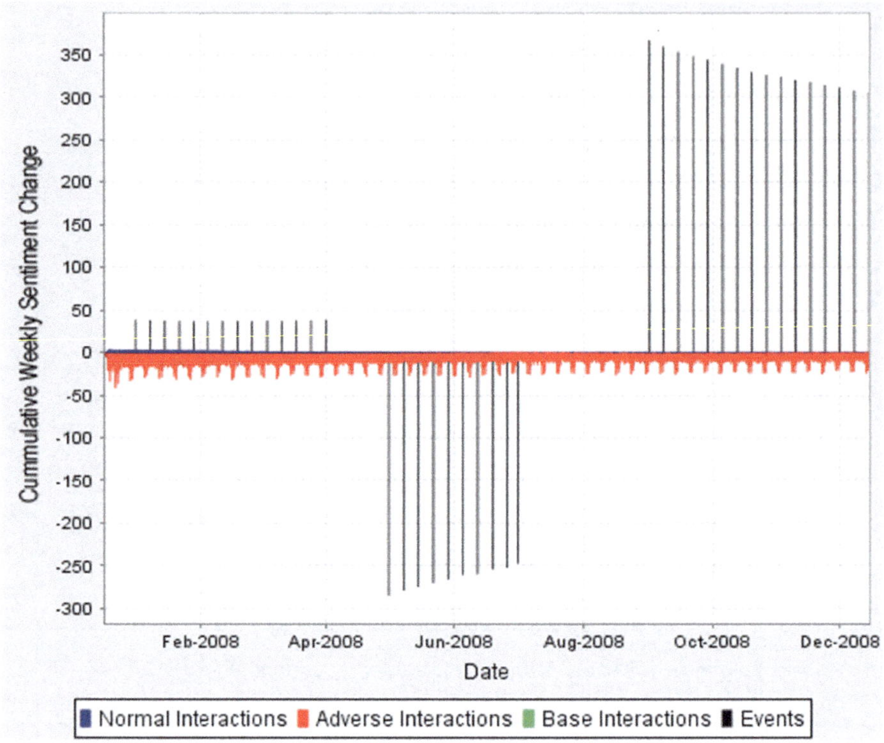

Fig. 10 Sentiment change by source

PMRFSim also examines information exchange. The information modeled by PMRFSim is called physical security information (PSI). Figure 11 shows both sentiment and information exchanges that result from non-PMRF personnel interacting with PMRF. These results are typical and show that most interactions result in a positive sentiment effect and PSI is randomly observed by agents visiting PMRF.

Figure 12 illustrates that many visiting agents that enter a restricted area (for any reason, including being lost) are not caught by PMRF security personnel. This is due to the normal force protection condition and relaxed levels of security used in this scenario. Figure 13 shows that after one year the adverse agent with the most PSI (i.e. the biggest single threat) has only acquired ~11% of PMRF's physical security integrity information (Fig. 14).

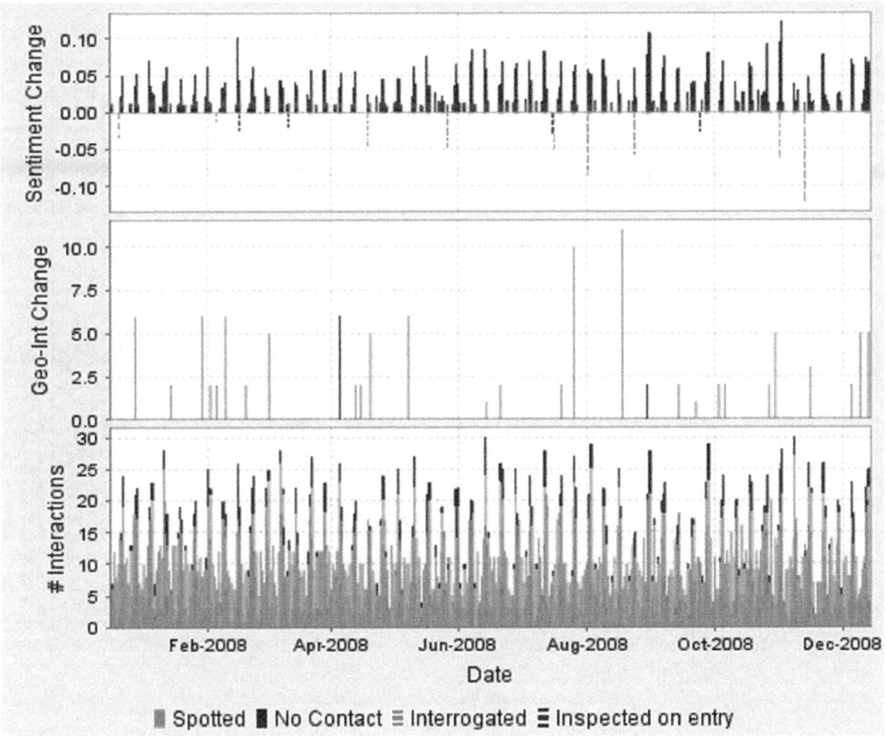

Fig. 11 Base interaction results

7 Conclusions

PMRFSim is a research-level social agent-based simulation capable of examining the interactions of 60,000+ agents over a simulated year within a few minutes. Many inputs are required from sources ranging from the U.S. Census Bureau, regional experts, and base security. Numerous graphs (and Google Earth playback videos) allow the analyst to get a feel for the social landscape that results from particular scenarios and initial assumptions. This analysis, in turn, feeds into decisions such as where and how to allocate public affairs efforts. PMRFSim provides a rudimentary capability that can be leveraged by a public affairs officer (PAO) to help assess, visualize, and explore the current socio-cultural situation, allowing the PAO to better target audiences and giving the PAO a clearer understanding of the public perceptions of the facility.

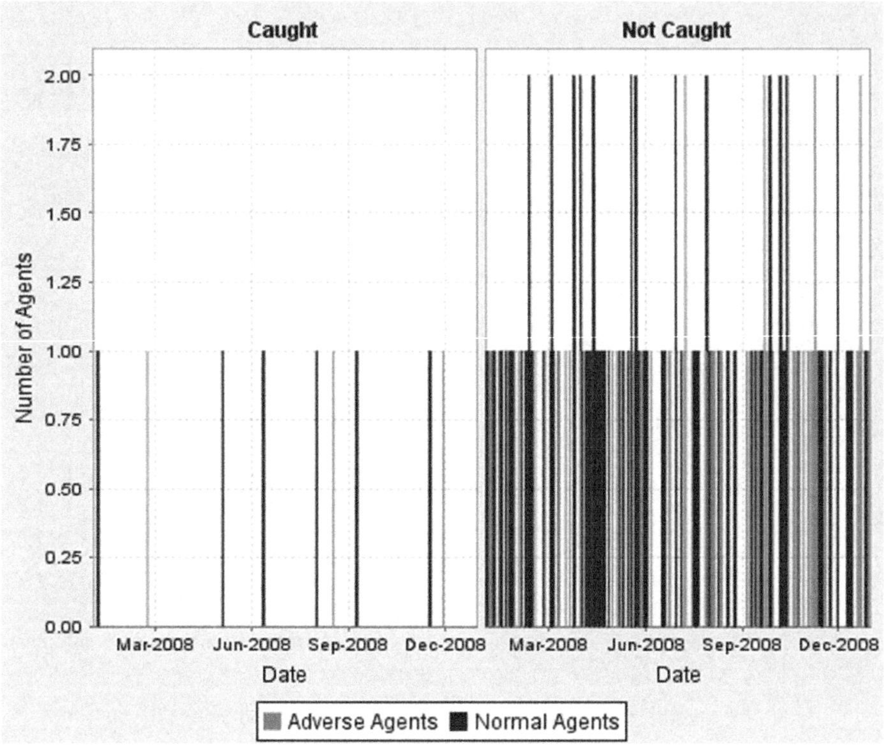

Fig. 12 Caught or not

PMRFSim is an example of an advanced analytic environment that supports not only the evaluation of the "communicable impact" of a given security strategy on the surrounding community, but provides a means of considering how an adversary may exploit locally held knowledge or perspectives about the facility. This latter capability speaks to how an adversary may develop the "capacity to act" by interacting with the local community.

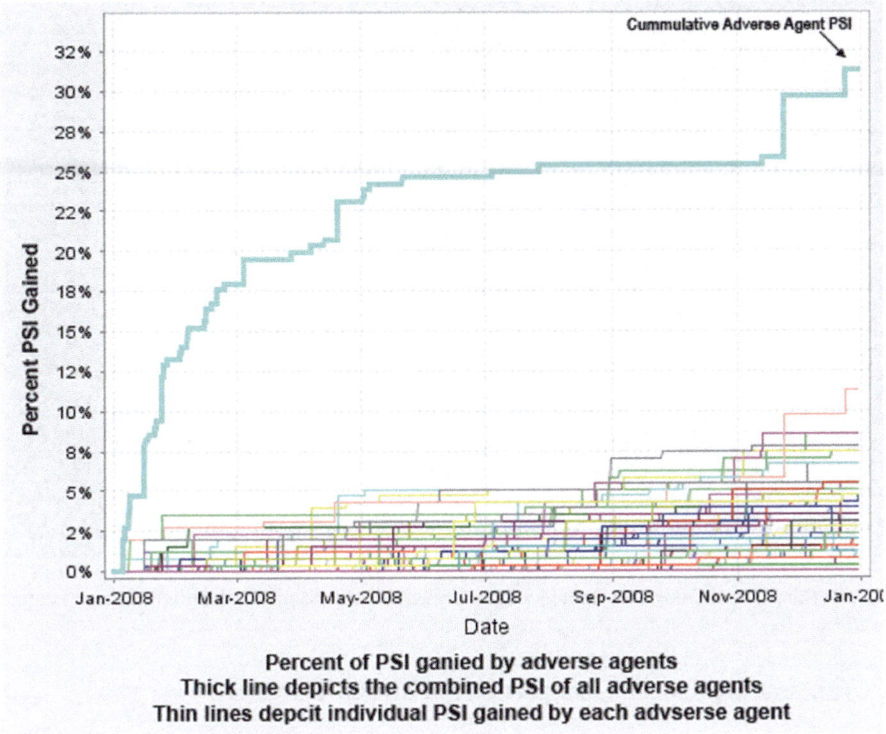

Fig. 13 Compromised intelligence

Alternatively, it is also possible to use members of the community as both sensors (they identify signals by the types of questions they are asked) and counter-sentiment projectors (they provide counter-valing messages to antagonist or PMRF detractors). Thus, the environment allows for understanding the interrelationship between the notions of base protection strategy and strategic communication. We view the future exploration of this relationship, i.e. the use of "soft power" as a means of enhancing security and situational awareness as an important research area.

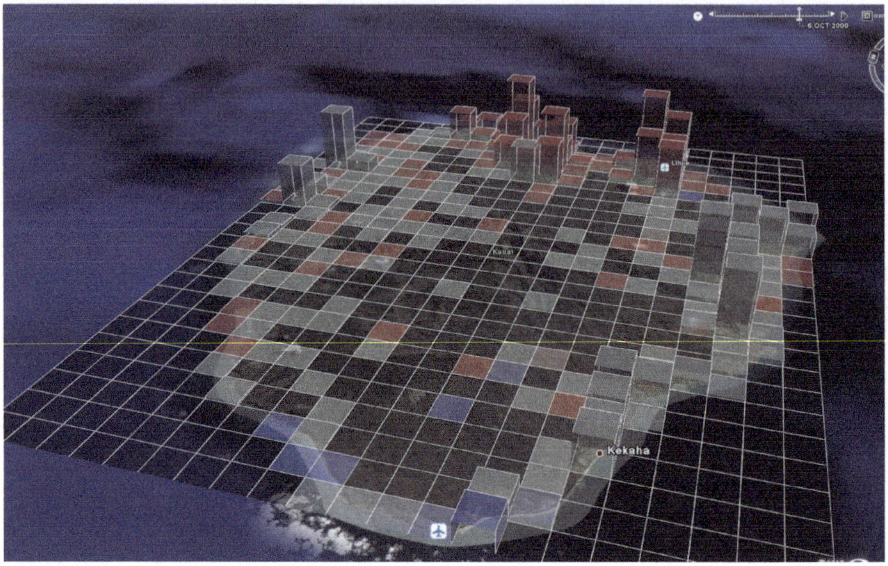

Fig. 14 Snapshot of Google Earth regional sentiment (color) and population (height) video output

8 Observations on Computational Social Science

The process of creating a computational social science (CSS) model involves defining, in computational terms, a complex social model that requires formally defining the environment, interactions, assumptions, and biases held by analysts and subject matter experts. It is the experience of the authors that this iterative process between the CSS team and the analysts is just as valuable to the analysts as any product derived from the final computational social science model itself. Analysts are forced to consider and deliberate over many questions to which the modeling team will require answers. Ultimately, the model is defined by the analysts and subject matter experts, so the final product is a visualization of their vetted biases, assumptions, and any limitations inherent to a real-world CSS model (e.g. incomplete data, necessary algorithmic assumptions and simplifications, etc.).

CSS is far from being able to make reliable social predictions. The real value of CSS, as it stands, is in the process of creating the model and in examining the dynamic interactions between modeled agents and groups that may or may not have been foreseen by analysts.

Acknowledgements This work was supported by US office of the Naval Research (ONR Contract N00014-07-C-0544). The authors would like to thank Dr. Sun-Ki Chai of the Hawaii Social Science Research Group for conducting sentiment analysis and identifying the topics of interest for the population under study. The authors would also like to thank Brian Chock from BAE Systems, Hawaii for his valuable feedback. Last, the authors would also like to thank Dr. John Sciortino for allowing us to use NRL's Java agent-based modeling packaged dubbed SWARM. The views and conclusions in this paper are those of the authors and should not be interpreted as representing official policies of the Office of Naval Research.

References

1. Axelrod, R.: The dissemination of culture: a model with local convergence and global polarization. J. Confl. Resolut. **41**(2), 203–226 (1997)
2. Bernardo, J., Smith, A.: Bayesian Theory. Wiley, New York (1994)
3. Byrne, D.: An overview (and underview) of research and theory within the attraction paradigm. J. Soc. Pers. Relat.Ships **14**, 417–431 (1997)
4. Epstein, J.M., Axtell, R.L.: Growing Artificial Societies: Social Science from the Bottom Up. MIT Press, Cambridge (1996)
5. Epstein, J.M.: Modeling civil violence: an agent-based computational approach. Proc. Natl. Acad. Sci. **99**(3), 7243–7250 (2002)
6. Frank, A., Saunders-Newton, D.: Journey from analysis to inquiry: Technology and the transformation of counterterrorism analysis. In: Yen, J., Popp, R. (eds.) Emergent Information Technologies and Enabling Policies for Counter Terrorism, pp. 315–329. IEEE/Wiley Interscience (2006)
7. MacKerrow, E.P.: Understanding why—dissecting radical islamist terrorism with agent-based simulation. Los Alamos Science **28**, 184–191 (2003)
8. McPherson, M., Smith-Lovin, L., Cook, J.M.: Birds of a feather: homophily in social networks. Ann. Rev. Sociol. **27**, 415–444 (2001)
9. Moon, I., Carley, K.M.: Modeling and simulation of terrorist networks in social and geospatial dimensions. IEEE Intell. Syst. Spec. Issue Soc. Comput. **22**, 40–49 (2007)
10. Pynadath, D.V.: PsychSim: agent-based modeling of social interactions and influence. In: Proceedings of the International Conference on Cognitive Modeling, pp. 243–348. ICCM Press, Pittsburgh, Pennsylvania (2004)
11. Schelling, T.C.: Micromotives and Macrobehavior. W. W. Norton, New York (1978)
12. Weisbuch, G.: Social opinion dynamics. Econophysics and Sociophysics, vol. 10, pp. 339–366 (2006)

Agent-Based Behavior Precursor Model of Insider IT Sabotage

Erika G. Ardiles Cruz, John A. Sokolowski, Timothy Kroecker and Sachin Shetty

Abstract Insider IT sabotage can be defined as the use of information technology to cause harm to an organization or an individual. Behavioral precursors are usually observable during the evolution of the threat. These precursors include personal predispositions such as rule and policy violations or mental health disorders, expressed disgruntlement due to unmet expectations or stressful events experienced by highly skilled employees with access to administrate systems, networks, and data in the organization. This research uses an agent-based modeling and simulation approach for modeling behavior precursors of insider IT sabotage within an organization using a risk scale. The specific behavioral precursors include the individual's predisposition, disgruntlement, stress levels, technical skill levels and the level of access to the computer systems. The simulation provides a framework for exploring the emergence and development of insider IT sabotage within organizations for different turnover rates.

Keywords Human behavior · Behavior precursor · Turnover rate · Insider IT sabotage · Agent-based modeling and simulation

---

E. G. Ardiles Cruz (✉) · S. Shetty
Old Dominion University, Norfolk, VA 23529, USA
e-mail: egama002@odu.edu

S. Shetty
e-mail: sshetty@odu.edu

J. A. Sokolowski
Old Dominion University, Norfolk, VA 23529, USA
e-mail: JSokolow@odu.edu

T. Kroecker
Airforce Research Laboratory, Rome, NY 13440, USA
e-mail: timothy.kroecker@us.af.mil

T. Carmichael et al. (eds.), *Complex Adaptive Systems*, Understanding Complex Systems, https://doi.org/10.1007/978-3-030-20309-2_4

1 Introduction

According to the Computer Emergency Response Team (CERT) "An insider threat is a current or former employee, contractor, or other business partner who has or had authorized access to an organization's network, system, or data and intentionally misuses that access to negatively affect the confidentiality, integrity, or availability of the organization's information or information systems". CERT considers the insider's use of Information Technology (IT) to harm the organization or a specific individual as insider IT sabotage [1]. CERT research on insider IT sabotage [2] found that the cases of sabotage were technically sophisticated and resulted in substantial harm to people and organizations. CERT also found behavioral precursors that are usually observable in employees during the emergence and evolution of the threat such as personal predispositions, disgruntlement due to unmet expectations, stressful events experienced by highly skilled employees with privileged access to administrate systems, networks, and data in the organization.

An organization's information security policies play an important role for the protection of assets. Assets to be protected from a cyber-attack are data and information in storage, and processing and communication phases. According to CERT [2], most organizations record and track employee's access to their systems, data, or network and employees are aware of monitoring policies. However, managers will sometimes shortcut essential monitoring policies or let them erode over time in an effort to increase productivity and morale. This environment provides a lower perception of the risk of attacks. One measure of the organization's economic growth is the turnover rate. The turnover rate is the rate of involuntary (fired or laid-off) or voluntarily leaving employees and the total of employees in the organization in a twelve-month period.

This research proposes to explore if turnover rates are correlated with the risk posed by insiders. The modeling approach is to develop an agent-based model of insider IT sabotage using a risk scale. The model includes two types of agents: employees and supervisors as well as their relationships with each other within an organization. The employees' modeling includes behavioral precursors of predisposition, disgruntlement, stress, skill-level, and access, while the organization's modeling includes the turnover rate. Using simulation, the model provides a framework to analyze the emergence and evolution of insider IT sabotage for different turnover rates. The remainder of this paper is organized as follows: (2) background, (3) methodology, (4) simulation, and (5) Conclusions.

2 Background

Cybersecurity measures are focused on threats outside an organization rather than threats posed by employees inside an organization, but insiders have likely caused more damage to an organization than outsiders. According to the Computer Emergency Response Team (CERT) "An insider threat is a current or former employee, contractor, or other business partner who has or had authorized access to an organization's network, system, or data and intentionally misuses that access to negatively affect the confidentiality, integrity, or availability of the organization's information or information systems" [1]. These violations can be the result of carelessness or accidents, but most damage is caused by the deliberate and intended actions of employees such as illegal access to a system, network, theft, destruction of data, or the compromise of network configurations. When technical employees use their abilities to sabotage the organization's system, data, or network as revenge for negative work-related events, it can be considered insider IT sabotage, changing the focus from productive work to sabotage, theft, or specific harm toward an individual. Alternatively, when employees experience personal financial difficulties and use their access to obtain clients' personal information to commit fraud, the primary motive is personal monetary profits. In other cases, employees may be motivated by greed, revenge, desire to obtain a business advantage, or the wish to impress a new employer, steal confidential data, proprietary information, or intellectual property. In these cases, it is considered theft. The motives and benefits for each scenario are different as are the behavioral precursors that employees exhibit. This research focuses on the concerning behavior exhibited by employees prior to the insider IT sabotage.

2.1 Insider IT Sabotage Behavior

According to CERT "Insider IT sabotage crimes are those in which insiders use IT to direct specific harm to an organization or an individual" [1]. Research on insiders' behavior highlights the importance of recognizing and addressing external signs of concerning behavior [2–7]. According to the Electronic Monitoring & Surveillance Survey of the American Management Association 2008, 43% of companies monitor employee e-mail, 66% monitor internet connections, and 45% track content, keystrokes, and time spent at the keyboard. Over 58% of managers surveyed have fired workers for email or Internet misuse. Of the employees dismissed, 64% violated company policy and 22% breached confidentiality rules.

Models of insider threat developed using behavioral precursors, technical indicators, or system access are focused on determining the likelihood of an employee committing a threat. CERT developed a system dynamics model of insider IT sabotage that considers mitigation of disgruntlement, this research also proposed general observations about insider IT sabotage behavior based on comparisons between empirical data and study findings [2]. Greitzer and Kangas developed a Bayesian network probability model of insider threat [8] finding that low risk indicators such as lack of dependability and absenteeism when observed with high risk indicators such as disregard for authority, disgruntlement, and anger management increase the risk of an insider crime. Kandias and Mylonas considered psychological profiling and real time use profiling to develop a decision making model of insider threat based on a scoring system [9]. Sokolowski, Banks, and Dover developed an agent-based model of insider threat using Epstein's agent zero structure [10] that considers disgruntlement, rational behavior, and social contagion as the drivers of insider's behavior while organizational culture influences disgruntlement of employees [11]. Our approach uses a scoring system (low, medium, high) to measure the risk of insider IT sabotage based on the behavior exhibited by employees. This behavior considers predispositions, stress, and disgruntlement among others.

On the agent's level, we propose to represent insiders' behavior by accounting for five main characteristics found in the previous research: predispositions, disgruntlement, stress, skills and access using a risk scale to categorize the risk of the threat.

According to CERT [2, 3], predispositions include mental disorders, inappropriate behavior, poor socialization skills, decision making bias, and rule violations previous to the current job. CERT [2, 3] defined disgruntlement as the measure of an employee's unmet expectations in his current job position because of insufficient salary/bonus, lack of promotion, restriction of online access, limitations on use of company resources, violations of privacy in the workplace, diminished authority/responsibilities, perceived unfair work requirements, and poor coworker relations amongst others. Disgruntlement was modeled as the difference between expectations and fulfillment of expectations [2, 3, 12]. Employee stress can be defined as a state of mental or emotional strain resulting from adverse or demanding circumstances. CERT modeled stressful events such as calls of attention per poor performance, unacceptable behavior, excessive absenteeism, or violations of internet access [2, 3]. Technical skills are defined as the knowledge and abilities to accomplish technical tasks, for this specific model we consider that skills are acquired while employees perform their tasks. Access policies ensures access to resources and users. User authorization is granted when an employee is hired and provides access paths to the system, network, or data. The security level is granted to resources (objects) and allows users to access resources to perform tasks with five access attributes: read access, append access, execute access, write access, and control access. According to Bell and LaPadula principle of least privilege, users should be given the minimum authorization required to perform their tasks, and resources should be given the maximum security level that allows to perform the tasks requiring the resource [13–15].

We use agent-based modeling to represent dynamic changes over time for these five characteristics due to interactions between agents (supervisors, employees) with each other and their environment as insider IT sabotage emerges and evolves.

2.2 Turnover Rates

According to the Bureau of Labor Statistics, the turnover rate is the number of total separations both voluntarily and involuntarily, divided by average employment for the year. Turnover rates are related to the economic growth of an organization, industry, region or government. While organizations try to increase productivity by implementing policies that attempt to reduce their turnover rates or decreasing costs, employees tend to seek out better work incentives such as higher salaries, quality health care plans, and retirement plans. Huselid evaluated the links between systems of high performance work practices and firm performance. His results indicated that these practices have an economically and statistically significant impact on turnover and productivity [16]. Guthrie found a positive association between use of high-involvement work practices, employee retention, and firm productivity [17]. On the organization level, our approach considers using simulation to explore the impact of turnover rates on the risk of a threat to assess the hypothesis that turnover rates are correlated with insider IT sabotage risk.

3 Methodology

One modeling approach suitable for representing complex human systems is agent-based modeling and simulation. The use of an agent-based model allows us to model human behavior within a complex human system representing agents and their interactions with each other, and with the environment according to simple behavioral rules in a virtual computer environment. While fundamental social structures and group behavior emerge from these interactions, the simulation analysis will lead to determine which factors have the highest degree of influence on the outcome.

The following components were developed in this section: the environment setting, agents' attributes and initial values, agent's behavior for updating dynamic attributes, the risk scale setting, and the verification and validation of the model.

3.1 The Environment

The environment (organization) is represented by a grid (39 × 39 squares). The model was developed in NetLogo 5.0.5 agent based programmable modeling environment for simulating natural and social phenomena [18]. The organization has two types of agents: employees and supervisors, and the behavior for rules that drive agents' interactions with each other and their environment. Initially we set the number of employees to 500 and the number of supervisors as 2% of the number of employees. Employees can leave the organization voluntarily or involuntarily, and the organization must hire and train a new employee for this position. This rate of change is called turnover rate *(t)* and for this model the turnover rate is set to 15%.

3.2 The Agents

Supervisors. Supervisors have the following attributes: position (x, y) on the grid initialized randomly and local vision in a Von Neumann configuration (4 neighboring

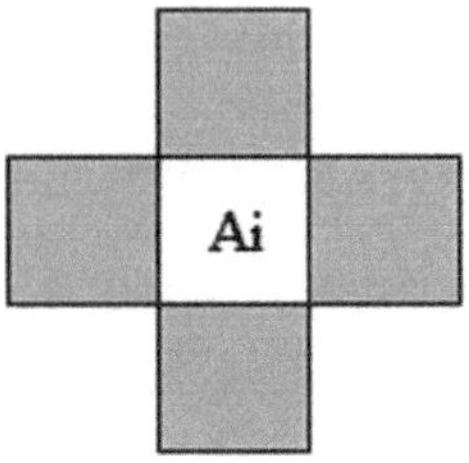

Fig. 1 Local Vision for each agent colored gray

cells). Figure 1 below shows the agent's local vision from which they can collect information.

Employees. Employees have the following attributes: position (x, y) on the grid initialized randomly, local vision in a Von Neumann configuration (4 neighboring cells), and insider IT sabotage risk (R) calculated as the summation of the five following attributes: personal predispositions (P), disgruntlement due unmet expectations (D), level of stress (S), level of skills (K), and level of access (A). These five attributes are measured between 0 and 1, where 0 represents the lowest possible value and 1 the highest possible value.

Personal Predispositions (P). This attribute measures indicators of mental disorders, inappropriate behavior, poor socialization skills, decision making bias, or rule violations previous to the current job position. Predisposition is set at 0 for employees without predispositions and 1 for employees with predispositions. For this specific model, predisposition is set with a Bernoulli probability distribution with 0.1 probability.

Disgruntlement (D). This attribute measures an employee's unmet expectations in his current job position such as insufficient salary/bonus, lack of promotion, restriction of online access, limitations on use of company resources, violations of privacy in the workplace, diminished authority/responsibilities, perceived unfair work requirements, poor co-worker relations. Disgruntlement is set initially on 0. Disgruntlement is a dynamic attribute that will be updated each time tick as the difference between expectations and fulfillment of expectations [2, 3, 11].

Stress (S). This attribute represents stressful events such as calls of attention per poor performance, unacceptable behavior, excessive absenteeism, or violations of internet access. Stress is set initially on 0. Stress is a dynamic attribute that will be updated each time tick based on the Rescorla Wagner Model [19].

Skills (K). This attribute assesses the skills acquired by the employee since the beginning of the current job position. Skills is set heterogeneous and uniformly distributed for all employees with a Uniform (0,1) probability distribution. Skills is a dynamic attribute that will be updated at each time tick.

Access (A). This attribute measures the level of access to the network, information, or information systems. It is set to 0 for no access, 0.5 for user privilege, and 1 for administrator privilege. For this specific model, 10% of employees have Access set

Table 1 Agent's attributes, initial configuration and behavior rules

Agent	Supervisor	Employee
Attributes and Initial Configuration	x = Random	x = Random
	y = Random	y = Random
	Vision= neighborhood4	Vision = neighborhood4
		Predisposition (P) = Bernoulli(0.1)
		Disgruntlement (D) = 0
		Stress (S) = 0
		Skills (K) = Uniform(0,1)
		Access(A) = $\begin{cases} 0.0 & 0.0 \leq \text{Uniform}(0,1) \leq 0.1 \\ 0.5 & 0.1 < \text{Uniform}(0,1) \leq 0.9 \\ 1.0 & 0.9 < \text{Uniform}(0,1) \leq 1.0 \end{cases}$
		Insider Threat Risk (R) = P + D + S + K + A
Behavior	Move	Move
		Update_Disgruntlement
		Update_Stress
		Update_Skills
		Hire_Employee
		Update_Risk

to 0, 80% of employees have access set to 0.5, and the remaining 10% have access set to 1. Access is a static attribute that does not change over time.

Insider Threat Risk (R). Finally, the insider IT sabotage risk attribute is the summation of all the before listed attributes.

Table 1 below summarize agent's configuration, initial values, and behavior rules developed in Sect. 3.3 for employees.

3.3 Behavior Rules

Move. Move is a behavior rule implemented for all agents to allow them to displace within the grid, in each simulation time, all agents select randomly a location in their local vision and displace to it. Move behavior rule allows employees to get updated information in each time tick from supervisors within their local vision to update the attribute stress.

Update Disgruntlement. Update Disgruntlement behavior rule is implemented for predisposed employees. In each simulation time, predisposed employees will update disgruntlement as the difference between insider's expectations and fulfillment of those expectations [2]. Employees' *Fulfillment (F)* represents how insider interprets the organization's actual ability or failure to meet expectations as shown in Eq. 1. It is set with a normal probability distribution with mean organizational culture and variance organizational change, for this model set at 0.8 and 0.05 respectively.

$$F_i = normal(org\_culture, org\_change) \tag{1}$$

An insider's *Expectations (E)* change based on initial *(F_1)*, current *(Fi)*, and historical fulfillment of those expectations as shown in Eq. 2. The primacy weight *(p)* represents how much weight the employee places on his initial fulfillment, the consistency weight *(c)* represents how much weight the employee places on his average fulfillment over time, and the recency weight *(r)* represents how much weight the employee places on his most current fulfillment. For this model, p, c, and r are set to 1.

$$E_{i+1} = \frac{p \cdots F_1 + c.\frac{\sum_{j=1}^{i} F_j}{j} + r.F_i}{p + c + r} \tag{2}$$

The agent's interpretation of fulfillment deficit (disgruntlement) is calculated as the difference between expectations and fulfillment of expectations as shown in Eq. 3 below. The agent specific affective weight a is set to 1. The model of disgruntlement and the initial configurations of parameters were set based on the insider threat agent based model developed by Sokolowski, Banks, and Dover [12].

$$D_i = (E_i - F_i).a \tag{3}$$

Update Stress. Update Stress behavior rule is implemented for employees. Increase stress is an event driven behavior rule. Every time an explore meets a supervisor within his local vision, his stress level will increase according to the Rescorla Wagner Model for acquisition of stimuli [19] with leaning rate α_1 set on 0.5 and maximum level of stress λ_1 set to 1. As shown in Eq. 4 below.

$$S_i = S_{i-1} + \Delta S \tag{4}$$

$$where \quad \Delta S = \alpha_1.(\lambda_1 - S_{i-1})$$

After the stimuli is no longer present, stress will decrease on time based on the Rescorla Wagner Model for extinction of stimuli [19] with extinction rate α_2 set on 0.25 and minimum level of stress λ_2 set to 0. As shown in Eq. 5 below.

$$S_i = S_{i-1} - \Delta S \tag{5}$$

$$where \quad \Delta S = \alpha_2.(\lambda_2 - S_{i-1})$$

Table 2 Insider threat sabotage risk scale

Risk value	Set color to	Insider IT sabotage risk category
0.0–2.5	White	No risk
2.5–3.0	Green	Low
3.0–3.5	Yellow	Medium
3.5–5.0	Red	High

Update Skills. Update Skills behavior rule is implemented for employees. CERT research on insider IT sabotage found that threats were performed mainly by highly skilled employees with privileged access [2]. We assume that employees acquire these technical skills to perform their tasks as time evolves in their current job position. In each simulation time, all employees will increase skills heterogeneously according to a normal probability distribution with mean 0.001 and variance 0.0005. As shown in Eq. 6 below.

$$K_i = K_{i-1} + normal(0.001, 0.0005) \tag{6}$$

Hire Employee. Hire Employee behavior rule is implemented for employees. In each simulation time, if an employee leaves voluntarily or involuntarily with probability equal to the organization's turnover rate (t), then a new employee will be hired. In this case, the attributes disgruntlement (D), stress (S), and skills (K) are set all on 0, while the attribute access (A) keeps the same because the new employee will be hired for the same position rather than the leaving employee and predisposition (P) will be set random with a Binomial(0.1) probability distribution.

Update Risk. Update Risk is a behavior rule implemented for employees. In each simulation time, the risk of an employee committing insider IT sabotage is calculated as the summation of the five attributes predisposition, disgruntlement, stress, skills, and access. As shown in Eq. 7 below.

$$R = P + D + S + K + A \tag{7}$$

Then each employee is color coded according to a risk scale to categorize the risk of an employee committing insider IT sabotage from lower to higher. This color scale allows the collection of statistics about the number of employees with low, medium, and high risk to commit insider IT sabotage. It is important to notice that different organizations can adjust this risk scale according to their own policies and expectations to decrease the risk of insider IT sabotage risk. Table 2 below summarizes the risk scale used for the current model.

3.4 Verification and Validation

The verification and validation of the model was done by analyzing model outputs with subject matter experts, looking for results that showed reasonable behavior, the repeatability of the experiment, and ITS risk values within the domain of definition. Future research should be done to address the validation of the current model with qualitative and/or aggregated empirical data.

4 Simulation

The simulation was developed using a mono factorial experimental design for analyzing the effect of turnover rates *(t)* on the insider IT sabotage risk *(R)* within the organization. Simulation time was set to 365 days. We performed 30 replications of 120 runs, setting turnover rates randomly distributed between 10 and 60%. The turnover rates settings were based on the statistics collected by the Bureau of Labor and Statistics for Annual Turnover Rates 2012 by Industry (https://www.bls.gov/bls/news-release/jolts.htm#2016), this report has a mean turnover rate of 37.4% and turnover rates varying from 14.0 to 55.8% for industries such as Mining and Lodging, Manufacturing, Trade Transportation and Utilities, Information, Financial Activities, Professional and Business Services, Education, Healthcare, Government among others. We recorded the average of employees in each category of insider IT sabotage risk (low, medium, high), and performed a correlation analysis. Figure 2 below summarizes results.

The plot shows three categories of insider IT sabotage risk. The green dots represent low insider IT sabotage risk, the yellow dots represent medium insider IT sabotage risk, and the red dots represent high insider IT sabotage risk. For employees with low risk of insider IT sabotage the correlation coefficient is − 0.98 and slope of −0.066. For employees with medium risk of insider IT sabotage the correlation coefficient is −0.99 and slope of −0.064. Finally, for employees with high risk of insider IT sabotage the correlation coefficient is −0.95 and slope of −0.014. The simulation results shown an inverse correlation between risk of insider IT sabotage and turnover rates for low, medium, and high categories of risk. Simulation results also shown a bigger rate of change for employees with low risk of insider IT sabotage than for employees with high risk of insider IT sabotage.

From the modeling point of view, the turnover rate represents the number of employees leaving per year within the organization. When an employee leaves (no matter the insider IT risk category), he will be replaced with a new employee that has no skills, no stress, no disgruntlement, the same access, and can be or not be predisposed. We believe that the insider IT risk (R = Predisposition + Disgruntlement + Skills + Stress + Access) on average is lower for the newly hired employee than for the departing employee. This is the mechanism that allows for decreasing the risk of insider IT sabotage.

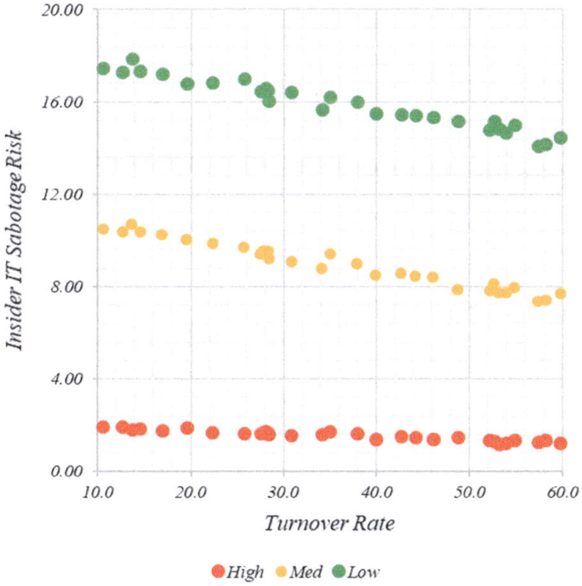

Fig. 2 Insider IT sabotage risk per category

Turnover rates are important indicators of profit for organizations. High turnover rates can decrease productivity and increase costs. Human Resource departments spend time and money advertising job openings, recruiting candidates, interviewing, and conducting new-hire training. The productivity of companies tends to slow down as the new employee adjust and learn the new tasks and responsibilities.

Research on insiders' behavior prior to the threat suggest that there was indication of concerning behavior when disgruntlement arose. However, the concerning behavior was not addressed and it evolved into a negative behavior or sabotage mainly performed by former employees. The research also provides possible ways to mitigate the risks to threat/sabotage by providing healthier workplace environments. CERT [2] proposed mitigation of disgruntlement, and expectations setting. Willinson and Warkentin addressed mitigation of disgruntlement by organizational justice [20]. Sokolowski, Banks and Dover found that no insider threats were likely to develop within a favorable organizational climate [12].

5 Conclusions

In this research, agent-based modeling and simulation was used for modeling and analyzing complex adaptive human behavior within an organization. In the current model, the local interactions of employees with each other and the environment have an impact on the emergence and evolution of insider IT sabotage. The simulation results show the inverse correlation between turnover rates and insider IT sabotage for the three categories of insider IT sabotage risk (low, medium, high).

Preliminary results support research on improving workplace environments to reduce the risk of a threat. Organizational policies must be developed to encourage a healthy work environment in which all employees feel treated fairly by the same rules. In addition, they should encourage a trusting workplace environment, in which counseling services can help employees to adapt and overcome the challenges that the new or current job can produce [2, 12, 20].

Improvements to the current model can include the verification and validation of the current model based on the availability of qualitative and/or aggregated empirical data of insider IT sabotage. Future work could include modeling technical precursors of a threat such as system access count, system files download, and hacking tools download. Finally, future work can also include the development of an optimization model for minimizing the risk of insider IT sabotage based on the behavior precursors developed in this research.

Acknowledgements This work is supported by the Office of the Assistant Secretary of Defense for Research and Engineering (OASD (R&E)) agreement FA8750-15-2-0120

Disclaimer

The views and conclusions contained in this paper are those of the authors and should not be interpreted as the as necessarily representing the official policies or endorsements the Office of the Assistant Secretary of Defense for Research and Engineering (OASD (R&E))

References

1. Cappelli, D.M., Moore, A.P., Trzeciak, R.F.: The CERT Guide to Insider Threats: How to Prevent, Detect, and Respond to Information Technology Crimes (Theft, Sabotage, Fraud). Addison-Wesle (2012)
2. Moore, A.P., Cappelli, D.M., Trzeciak, R.F.: The "big picture" of insider IT sabotage across US critical infrastructures. In: Insider Attack and Cyber Security, pp. 17–52. Springer, (2008)
3. Band, S.R., Cappelli, D.M., Fischer, L.F., Moore, A.P., Shaw, E.D., Trzeciak, R.F.: Comparing insider IT sabotage and espionage: a model-based analysis. DTIC Document (2006)
4. Colwill, C.: Human factors in information security: the insider threat–Who can you trust these days? Inf. Secur. Tech. Rep. **14**(4), 186–196 (2009)
5. Crossler, R.E., Johnston, A.C., Lowry, P.B., Hu, Q., Warkentin, M., Baskerville, R.: Future directions for behavioral information security research. Comput. Secur. **32**, 90–101 (2013)
6. Greitzer, F.L., Kangas, L.J., Noonan, C.F., Dalton, A.C., Hohimer, R.E.: Identifying at-risk employees: modeling psychosocial precursors of potential insider threats. In: 2012 45th Hawaii International Conference on System Science (HICSS), pp. 2392–2401. IEEE
7. Shaw, E.D., Fischer, L.F., Rose, A.E.: Insider risk evaluation and audit. DTIC Document (2009)
8. Greitzer, F.L., Kangas, L.J., Noonan, C.F., Dalton, A., Hohimer, R.: Identifying at-risk employees: a behavioral model for predicting potential insider threats. Pacific Northwest National Laboratory Richland, WA (2010)
9. Kandias, M., Mylonas, A., Virvilis, N., Theoharidou, M., Gritzalis, D.: An insider threat prediction model. In: International Conference on Trust, Privacy and Security in Digital Business, pp. 26–37. Springer (2010)
10. Epstein, J.M.: Agent_Zero: Toward Neurocognitive Foundations for Generative Social Science. Princeton University Press, Princeton (2014)

11. Sokolowski, J.A., Banks, C.M.: Agent implementation for modeling insider threat. In: 2015 Winter Simulation Conference (WSC), pp. 266–275. IEEE
12. Sokolowski, J.A., Banks, C.M., Dover, T.J.: An agent-based approach to modeling insider threat. Comput. Math. Organ. Theory **22**(3), 273–287 (2016)
13. Bell, D.E., LaPadula, L.J.: Secure computer systems: mathematical foundations. DTIC Document (1973)
14. LaPadula, L.J., Bell, D.E.: MITRE technical report 2547, volume II. J. Comput. Secur. **4**(2–3), 239–263 (1996)
15. LaPadula, L.J., Bell, D.E.: Secure computer systems: a mathematical model. Technical report 25471996
16. Huselid, M.A.: The impact of human resource management practices on turnover, productivity, and corporate financial performance. Acad. Manag. J. **38**(3), 635–672 (1995)
17. Guthrie, J.P.: High-involvement work practices, turnover, and productivity: evidence from New Zealand. Acad. Manag. J. **44**(1), 180–190 (2001)
18. Wilensky, U.: "Netlogo, 1999," Center for Connected Learning and Computer-Based Modeling. Northwestern University, Evanston, IL (2010)
19. Rescorla, R.A., Wagner, A.R.: A theory of Pavlovian conditioning: variations in the effectiveness of reinforcement and nonreinforcement
20. Willison, R., Warkentin, M.: Motivations for employee computer crime: understanding and addressing workplace disgruntlement through the application of organisational justice. In: Proceedings of the IFIP TC8 International Workshop on Information Systems Security Research. International Federation for Information Processing, pp. 127–144 (2009)

Formal Measures of Dynamical Properties: Tipping Points, Robustness, and Sustainability

Aaron Bramson

Abstract Complex systems are characterized by processes that exhibit feedback, nonlinearity, heterogeneity, and path dependencies, and accurately modeling such systems is becoming increasing important. To help realize the potential of complex systems modeling we need new methods that are capable of capturing the dynamical properties of such processes across disciplines and modeling frameworks. This chapter presents a portion of the methodology development that includes formal and domain-agnostic definitions of phenomena related to tipping points, criticality, robustness, and sustainability. For each included concept I provide a probabilistic definition based on a Markov model generated from time-series data in a specific way. These rigorous mathematical definitions clearly distinguish multiple distinct dynamical properties related to each concept, and they also function as measures of these properties. Though only a small portion of the methodology's capabilities, theorems, and applications can be included in this treatment, it does include all the foundational material necessary to apply the methodology.

1 Introduction

The dynamics of system behavior exhibit a range of properties that must be better understood for us to predict, control, and explain phenomena across all disciplines. There are many challenges to face in addressing the identification and measurement of dynamical properties: the concepts must become well-defined, data must be captured

A. Bramson (✉)
Laboratory for Symbolic Cognitive Development, Center for Biosystems Dynamics Research, 2-1 Hirosawa, Wako City, Saitama 351-0198, Japan
e-mail: bramson@riken.jp

A. Bramson
Department of General Economics, Ghent University, Tweekerkenstraat 2, 9000 Ghent, Belgium

T. Carmichael et al. (eds.), *Complex Adaptive Systems*, Understanding Complex Systems, https://doi.org/10.1007/978-3-030-20309-2_5

and analyzed in a domain-agnostic way, and the measures must be interpretable for any (appropriate) application. The approach presented here combines modeling and analysis techniques from disparate fields to meet these challenges in a way that allows it to be fully general and widely applicable. With increasing concern about the robustness, sustainability, and tipping points of systems, the ability to detect, measure, and compare such phenomena within a mathematical framework becomes critically important.

Though the ideas of tipping points and robustness are gaining prevalence, their usage is informal and inconsistent. These dynamics are frequently tied up with processes such as feedback, externalities, continued operation, persistence, perturbation, etc., but not in ways that explicitly draw the connections. Specific models have used these concepts and sometimes provided idiosyncratic formal definitions and measures, but they are not general characterizations that can foster comparison across models, scales, disciplines, or applications.

This project aims to elucidate distinctions among dynamical properties based on features of system behavior that are independent of the substantive domain and hence of the generating mechanisms or process. Even the identification of tipping and robustness phenomena as features of *behavior* rather than of a generating mechanism is a substantive distinction. The definitions and concepts included here are not exhaustive; there are several more within this approach that are not included in this treatment, and many more that cannot be defined through this technique. It is important to keep in mind that the goal of a project like this is not to get the definitions of these terms right (as if there were a single correct definition), but instead to provide definitions that are methodologically and/or conceptual useful.

The conceptual distinctions are accomplished through the formal definitions put forth for each concept, and these definitions also function as measures the associated property. The mathematics involved utilizes a Markov model representation of time series data constructed in a particular way; what we call a *Revealed Dynamics Markov Model* (or RDMM). The details of creating the necessary Markov representation are explained in detail in other work (e.g. [2]) and will be only briefly covered here. With this mathematical foundation in place, probabilistic definitions for familiar landmarks in system dynamics (attractor, basin of attraction, support, etc.) and structural features (path, cycle, perimeter) are identified. These are leveraged to establish probabilistic definitions and measures for various tipping point and criticality-related concepts. All this material acts as a springboard for the definitions and measures of robustness-family concepts. To aid understanding, each definition is illustrated with simplified graphical examples that highlight the essential features of the proposed measure.

The chapter is organized in the following way. First, some background material is presented on the dynamical properties addressed by this methodology—this material contextualizes the methodology, but can be skipped by those primarily concerned with learning and applying the techniques. The technical material starts with the relevant features of Markov models and how to create one from time-series data. This is followed by definitions of the foundational concepts and landmark properties of system dynamics. After that the tipping point-related concepts are described, defined, measured, and illustrated. The next section builds on the earlier definitions to construct measures for the robustness-related concepts. The paper finishes with several branches of extensions: applications of these measures, more novel measures for dynamical properties, behavioral equivalence classes for system dynamics, and others.

2 Background and Motivations

My motivation for developing a general methodology for measuring dynamical properties springs from a need for better analyses of computational models (such as agent-based models of complex adaptive systems). Specifically, we lack appropriate techniques to (1) identify patterns in the behavior of non-equilibrium systems, (2) identify (and hence control) the features of mechanisms responsible for non-linearity, path dependence, and self-organization, and (3) understand processes in a non-aggregated causally-proximate way. Since dynamical properties are properties of *behavior*, they apply to the output data generated by such models rather than the generating mechanisms. As a result, this technique may be applied to any data set containing multiple, independent series of data points. And furthermore it can be used to encode and compare data sets generated through different means (e.g., for model porting, calibration, and verification with empirical data).

Much previous work in finding and measuring properties of system dynamics has focused on explanation—on answering the *why* questions. This is to be expected because these discussions were couched in scientific contexts where a particular phenomenon required explanation. Each such jaunt into explaining tipping points or robustness for a specific substantive problem was accompanied by a custom-suited methodology capable of generating and detecting that property *within the model provided* (to answer the *how* question). Thus each such measure was designed to explain the dynamical property within a particular model and domain. That approach is methodologically limiting because different models (even of the same phenomena) will have incommensurable measures of each property, and it may take considerable domain knowledge to interpret and understand each measure.

By contrast, the current work is one of pure methodology: It is meant to be completely abstract and general, and therefore capable of measuring these dynamical properties for any system (that satisfies the input data conditions). Thus the techniques presented below answer the *whether* and *how much* questions for dynamical properties—which are key to informing the *how* and *why* questions. A general methodology provides a framework through which all modelers (including data analysts) can determine whether and how much (for example) robustness obtains in their system of study … and compare results across models regardless of the generating mechanisms. The ability to compare measures across systems is achieved through a focus on scale-free measures—measures that do not depend on the size/resolution/dimensionality of the system being analyzed. In what follows I provide some additional background on each of the components of the methodology.

2.1 *Markov Modeling, Network Theory, Graph Theory, etc.*

The technique being presented renders multiple independent streams of time series data as a *Markov model*—a probabilistic state-transition diagram satisfying a specific set of mathematical conditions. This mathematical object is also a *network* of nodes and directed relationships (the states of the system and their transitions). It is also a *graph* à la graph theory when the states are considered as the vertices and the transitions as edges. Each of these fields has developed numerous methods for building, measuring, manipulating, and categorizing their respective version of this structure; many of which have been adopted and adapted for the current project.

Markov modeling has a long history in mathematics, engineering, and in applications to domains as diverse as condensed matter physics, genomics, sociology, and marketing. The use of Markov models to uncover information about system dynamics falls within the subfield of ergodic theory, and some of that material has been utilized here (with more to come). By using Markov models one can compute features such as the equilibrium distribution of states, expected number of steps between two states (with standard deviation), reversibility of processes, and periodicity of cyclical behavior. These mathematical results typically gain added meaning when interpreted for the substantive system being modeled, but this paper utilizes them as part of defining general properties of system dynamics.

Network theory, and especially the recent explosion in social network research, provides another suite of useful measures of network structure (and algorithms with which to calculate them). Common network properties such as node degree, diameter, betweenness centrality, clustering coefficients, and community structure can all be applied to the Markov model representation of data. The meaning of these measures in terms of system dynamics is not always clear, and there is still work to be done in exploiting available measures. It is also possible that novel measures developed for analyzing dynamical properties will feed back to network theory and provide new insights there.

Computer scientists have been analyzing physical networks in the form of actual communication networks and circuit boards, as well as various abstractions from these problems that now fall under the field of information theory. They have invented several useful measures (and exceptionally well-crafted algorithms) to calculate connectedness, load-bearing properties, path switching, transmission speed, and packet splitting/fusion to name a few. My analysis borrows heavily from this work in terms of algorithms for network flow, though each has been repurposed and reinterpreted for the abstract Markov model representation employed here. Hardware engineers and their physicist partners have worked out several interesting measures for circuit design problems: multiple paths, variable resistance, flow injection, capacitance, and many other characteristics of electronic circuits that also have analogs in the Markov models presented below.

And finally, graph theory offers a few useful measures for our purposes, and moreover provides a wealth of definitions for graph structure and node relationships. Structural properties (e.g., planarity, graph genus, projections, k-connectedness) will play a larger role in extensions addressing changes of resolution and in establishing equivalence classes of system dynamics (see Sect. 7). They may also play a useful role for particular applications of the technique in which system state properties coincide with features well-studied by graph theorists (e.g., discovering that a particular dynamical property obtains for sets of states that match the structure of regular k-partite graphs).

2.2 *Tipping Points and Related Concepts*

Tipping points (in their various versions) have captured the public's attention from topics as diverse as segregation, marketing, political unrest, material science, ecosystem stability, and climate change. However the concept of a tip has not been generally and formally defined and, as a result, the term's uses across these disparate applications are inconsistent. At times a tipping point refers to a threshold beyond which a system's outcome is known. Other times a tipping point describes an event that suffices to achieve a particular outcome, or an aspect of such an event, or the time of such an event. Another use of the term 'tipping point' is to label the conditions to which the system is most sensitive. And much too often the precise meaning is never specified.

The term 'tipping point' was first coined by Morton Grodzins in 1957 [9] to describe the threshold level of non-white occupants that a white neighborhood could have before "white flight" occurred. The term continued to be used in this context through the work of Wolf [15] and Schelling [13] who also extended the concept to other similar social phenomena. Though these researchers had a specific usage with narrow focus, the idea of a critical parameter value past which aggregate behavior is recognizably different spread across disciplines where its meaning and application varied considerably.

Malcolm Gladwell's pop sociology book *The Tipping Point* [7] seems to have played a significant role in bringing the term to the public's awareness, and is also likely related to its increased usage in scientific publications as well.[1] The notion of tipping point most frequently used by Gladwell is an event that makes something unusual (such as Hush Puppy shoes) become popular. More precisely this is a critical value for some parameter that suffices to produce a phase transition for percolation through certain heterogeneous social network structures. This form of tipping point behavior also appears in the work of Granovetter [8] and Young [16] for the propagation of rioting behavior and technology respectively. This version of tipping is often studied by physicists and is a well understood phenomena, though it will play only a minor role in what follows. However, the fact that the expression has made it into

[1] According to Google's Ngram viewer, the term "tipping point" appeared in English-language books (within Google's scanned corpus) steadily between 0.000002 and 0.000004% between 1976 and 2000 (although it saw a spike to 0.00001% in 1975). From 2000 to 2007 its usage increased steadily from 0.000004 to 0.000029%—more than a seven-fold increase.

the layman's conceptual vocabulary boosts the importance of establishing rigorous scientific definitions and usage.

Tipping points have also appeared in reports of climate change with increasing frequency. James E. Hansen has claimed that "Earth is approaching a tipping point that can be tilted, and only slightly at best, in its favor if global warming can be limited to less than one degree Celsius" [5]. This usage reflects Hansen's belief that "Humans now control the global climate, for better or worse." By contrast, Gabrielle Walker states, "A tipping point usually means the moment at which internal dynamics start to propel a change previously driven by external forces" [14]. It doesn't seem that Walker's and Hansen's comments about tipping points are compatible. And even if the distinct usages are meaningful within their fields, and useful for different purposes, they both fail as general characteristics. Identifying tipping points (as a property of system dynamics) should not depend on whether humans are in control of system behavior or what is driving the dynamics.

But not all heretofore definitions of the term 'tipping point' have been loose or subject-matter specific. It is often deployed as a semi-technical term in equation-based models of various sorts. For example, it can refer to an unstable manifold in a differential equation model, the set of boundary parameters for comparative statistics [12], or inflection points in the behavior of functional models. Since only a small proportion of systems bearing tipping behavioral properties can be readily and faithfully represented as systems of equations exhibiting these features, equation model-dependent definitions have limited usefulness. The approach presented below reveals the dynamical properties of any time-series data; it is not sensitive to how that data is generated. Though detecting certain properties or measure values certainly depends on the structure of the input data, the ability to apply the definitions/measures does not depend on the features input data.

2.3 Robustness Family Concepts

Robustness considerations are essential to understanding and maintaining systems of all kinds: engineered systems, ecologies, economies, political regimes, infrastructure, computer algorithms, homeostatic organisms, and decision procedures to name a few. There has been considerable growth in the number and variety of robustness measures due to recent trends in sustainable agriculture, climate change, economic development, social networking, and other nascent research thrusts.[2] Models in these fields are often developed specifically to enhance system robustness, avoid system failures, mitigate (or exploit) vulnerabilities, and otherwise cope with variable per-

[2]As a rough measure of the increased prevalence of robustness considerations, Google's Ngram viewer reports that the term "robustness" increases in usage in English-language books from 0.00002% in 1956 to more than 0.0002% in 2006 (with a slight drop since then)—a ten-fold increase. "Sustainable" quite remarkably goes from nearly 0.0 to 0.0014% over the same period (also dropping recently).

formance in an unpredictable environment—and in each one 'robustness' has it a domain/model-specific meaning and measure.

There is a wealth of terms associated with the maintenance of system properties: robustness, stability, resilience, reliability, sustainability, etc. Despite this variety, there are more distinct behavioral characteristics than there are terms, and no consistent mapping between them. I refer to this collection of concepts as the robustness family of concepts. Sometimes one wants to measure the likelihood that a property will continue to obtain, while at other times it is the ability to recover from failures, and sometimes it is how much of the time the property is present. These (and other) behaviors are all sometimes referred to as "robust", "reliable", "sustainable", etc., but are clearly distinct phenomena. The formal definitions and measures below pick out specific distinct system behaviors (matching distinct behavioral concepts) and associate them with terms that I think best match the nuance of the behavior. However there are more concepts than terms, and my focus is on the formal properties rather than the semantics of word choice.

Understanding how social systems can be both simultaneously flexible and strong has garnered increasing interest recently. Jenna Bednar describes how institutional design can affect the robustness of a federalist governing body [1] this way: "By explicitly acknowledging the context dependence of institutional performance, we can understand how safeguards intersect for a robust system: strong, flexible, and able to recover from internal errors." Bednar has identified the properties that make institutional systems robust (compliance, resilience, and adaptation) in a way that is somewhat specific to the subject matter. That is beneficial and to be expected for explaining and improving the robustness of political institutions. Such an analysis stands to gain from the conceptual refinements derived from formal measures of multiple robustness-related features of system dynamics. Bednar's work especially underlies the thought that understanding many systems of interest requires more than traditional equilibrium analysis; the dynamic nature of dissipative structures (see Example 6.5) requires new notions of stability, resilience, and robustness that I hope to help inform through the provided measures.

Thomas Sargent has made extensive use of principles from robust control theory in his analysis of monetary policy and pricing (and other topics). The sense of 'robustness' used in robust control theory is a gap between modeled levels and actual levels of parameters. It is used to formalize misinformation, uncertainty, and lack of confidence in agents' knowledge and, more generally, to facilitate high levels of performance despite errors and in situations known to be in less-than-ideal conditions. This sense of robustness applies across the decision theoretic sciences and planning literature [10]. However, not all robustness analyses are to cope with uncertainty.

In genetics the term robustness refers to a species' consistency of phenotype through changes in the genotype. Robustness can be considered at two levels: (1) how much mutation can a member of a species remain viable through and (2) how much genetic variation is required to transform a species' physical characteristics. The first level takes genetic profiles of organisms and from them determines which can survive to reproductive age and which cannot (or are sterile). The number of genotypic variations that remain viable is a measure of the species' robustness …

according to that usage. On the evolutionary time scale one may wish to understand how incremental genetic drift is responsible for large phenotypic variations over time. Walter Fontana has demonstrated that a network of neutral mutations (ones that do not affect fitness) can sufficiently explain the observed punctuated equilibria (see Example 6.4) in species evolution [6]. Though fitness (and even phenotype) may remain unchanged through some genetic variation, the connection between fitness change and phenotype change is strong. A model that tracks fitness through genetic variations could then approximately measure how robust each stage in the evolutionary progression is. It is clear that these two concepts of robustness are distinct; and they are also both distinct from the usages in control theory and decision theory.

We can add robustness measures from statistics and computer science to the variety of senses that 'robustness' can take. In computer science an algorithm, procedure, measure, or process is robust if small changes (errors, abnormalities, variations, or adjustments) have a proportionally small affect on the algorithm, procedure, measure, or process. The time complexity of two algorithms (for example) may change in different ways. Algorithm A may require one step per input ($O(n)$) and Algorithm B may require one step per two to the power of the input size ($O(2^n)$); in this case algorithm A is more robust to changes in input size. Statistical robustness is either when (1) an estimator performs sufficiently well despite the assumptions required by that estimator being violated or (2) when (like in computer science) a measure changes little compared to changes in the input. For example, the median is a more robust measure than the mean because to alter the median a data point has to cross the median point, whereas any input value change will change the mean's value.

And there are more variations: information theory, data security, engineering, law, ecology, and just about every field has their own versions of robustness. They all share certain high-level conceptual commonalities, but differ in their details and criterion for application. The definitions below yield necessary and sufficient conditions for the application of several robustness-related concepts—including some of those just described. The current technique does not capture every useful behavioral property in the robustness family, but it does succeed in capturing several key properties for understanding system dynamics and comparing them across domains. The level of conceptual refinement provided here may be able to bring many discussions of robustness from the different fields to a single table, and hence foster more fruitful interdisciplinary research and a general theory of robustness.

3 The Markov Model Representation

To fully grasp the definitions of system dynamics presented herein, one must understand how time-series data is used to create the Markov model representation. Fundamentally a Markov model is simply a collection of states and a set of transition probabilities between pairs of states. Different applications of Markov models take different system features as the states, but the nodes in the Markov models used for this methodology represent a complete description of an observed state of the system (details below). Given that states and transitions are defined this way, and since this analysis utilizes the Markov model as a *representation of the actually observed data*, it will be constructed as a static, finite and discrete-space, time-homogeneous Markov model. Before getting into the details, note that most computer simulation output already satisfies this data format and longitudinal data (e.g., census data, health records, and financial records) satisfy everything except time-homogeneity.

For this presentation, only a brief explanation is provided for the most common type of time-series data: homogeneous discrete-time data with real-valued and/or categorical state variables. The methodology is not limited to this kind of data; discrete-event and continuous-time processes can also be analyzed in an analogous way...and the probabilistic definitions provided later apply unchanged. Furthermore, though the conversion presented here coarse-grains values to generate a more manageable discrete-space transition diagram, and most users will benefit from such binning, it is also (theoretically) possible to run a modified analysis using a multi-dimensional hybrid vector/probability density space that does not require binning (although I have not yet attempted such an analysis).

Though the current treatment only presents enough details of the Markov model construction for the reader to understand the measures of dynamical properties, a more detailed account can be found elsewhere (e.g., [2] and future work). The analysis software under development includes sophisticated algorithms to generate this structure automatically according to options such as maximal sensitivity to a specified system property, maintaining a minimal confidence level for transition probabilities, or scale-matching for comparative analysis. In these cases, various forms of adaptive binning are utilized to cut up the space in the most useful way.

3.1 Constructing a Markov Model

As it is used here, a *system state* is simply a list of (quantitative or categorical) values that are considered to obtain concurrently. Typically this is a description of a time-slice of one run of a model's output data or all the collected data for one period of one trial of a dataset. We can consider a concurrent specification for each aspect of the system as an observed *configuration* of that system.

Definition 1 A *state* in the Markov model is a concurrently observed set of aspects (e.g., variables, parameters, system measures) of the system being analyzed. For a system in which each state has Q aspects,

$$S_i = \{X_{1(i)}, X_{2(i)}, \dots X_{Q(i)}\}$$

where $X_{h(i)}$ is the value of aspect h in state i.

For example, if we are analyzing health records, then we would take each individual's medical history as a trial. All the medical data (such as weight, height, blood pressure, and medications used, i.e., not including personal data that is unique to each individual.) recorded at each visit will comprise one state of the system. Note that the same list of values—the same configuration—may obtain at multiple periods within a trial and in multiple trials. In fact, the sharing of states across multiple trials is essential to the *usefulness* of this methodology.

Definition 2 Multiple observed specifications of aspects are represented as the *same state* of the Markov model if all the aspects of the two states are identically valued; i.e., distinct occurrences of the same configuration. For any pairwise comparison,

$$S_i = S_j \Leftrightarrow \forall h \; X_{h(i)} = X_{h(j)}$$

An obvious (but still useful) corollary results directly from the truth values of the biconditional equivalence used in defining identical pairs of states.

Corollary 3.1 *A difference in any aspect suffices to differentiate states of the system.*

$$\exists h \; X_{h(i)} \neq X_{h(j)} \Leftrightarrow S_i \neq S_j$$

Now let's look at a simple example of capturing the state-space of a formal model.

Example 3.1 Let's take as our system an iterated strategic-form (simultaneous move) game played by three players

$$P_j \in \{P1, P2, P3\}$$

each with four possible actions

$$a(P_j) \in \{A, B, C, D\}.$$

If we describe the system using only these actions, then each state of the system has three aspects (an action for each player) and each aspect takes on one of four values (the specific action taken by each one). That is

$$S_i = \{a(P1_{(i)}), a(P2_{(i)}), a(P3_{(i)})\}$$

and a particular state S_3 might be $\{C, B, C\}$. There are $3^4 = 81$ combinations of four actions for three players, each identifying a system configuration, but the Markov model may not include all of them. Recall that the Markov model is expected to be built from collected data, so some combinations of aspect values may be unobserved (and potentially theoretically impossible or simply irrelevant).

Observation 3.1 *Note that the number of nodes in the Markov model of your data is always less than or equal to the number of samples in your dataset (and usually much less than due to binning). So even if the potential state space is astronomically large, the generated Markov model is always manageable (or, at least as manageable as your data).*

For each trial of data, there is a sequence of specifications for every aspect of the system, and the list of aspects must be the same for each state. So, in the previous example, if it is possible for a player to leave the game, then that must correspond to an included value of $a(P_j)$—it cannot simply be omitted from the configuration. Each configuration corresponds to a unique node in the graph representation. A set of n states is demarcated with boldface type: $\mathbf{S} = \{S_1, S_2, \ldots S_n\}$. The set of all the states in the Markov model is $\mathbf{N}$ which has size $|\mathbf{N}| = N$; thus N is also the number of nodes in the graphical representation.

The state of the system at time t (denoted s_t) changes to s_{t+1} in discrete, homogeneous time intervals.[3] It is sometimes possible that $s_t = s_{t+1}$ either through a lack of activity or a transition to an indistinguishably different state (i.e., the values are all within the same bin). More generally it is possible for $s_t = s_{t+h}$ for any $h \geq 0$. In the Markov model's graphical form, a transition from S_i to S_j is represented as a weighted, directed edge from the node representing S_i to the node representing S_j.

We construct the Markov model representation through the following process (see Fig. 1). **Step 1**: Generate a node representing the first state in a time series; i.e, the first list of aspect (variable) values in the first trial. **Step 2**: For each subsequent *distinct* specification (set of values) in that sequence, we add another node. While doing this, we keep track of how frequently each *transition* between each pair of states has occurred (including returns to previous states and self-transitions). **Step 3**: Repeat this process for each sequence (trial) in the sample, continuing the accumulation of transition frequencies between each pair of states. **Step 4**: Using the normalized transition frequencies to assign probability weights for each state transition. **Step 5**: If any node lacks an exit transition, then create a self-transition with probability 1 and observed frequency 0.

Simulations often end when the simulation is no longer producing interesting data and some halting condition has been satisfied (which is often not an equilibrium condition). Thus some simulation data, and typical collected empirical data, may end on unique states. But because in a Markov model the outgoing probabilities must sum to one, we use step 5 to ensure a valid Markov model is created. The collection of frequency data can also be used to calculate confidence measures for those transition probabilities using an appropriate nonparametric multinomial method.

The final Markov model representation encodes state transitions as observed frequency-based probabilities that are captured as weights on the graph edges, and hence the graph's adjacency matrix doubles as Markov transition matrix. We write the probability of transitioning from state S_i to state S_j as $P_{ij} := P(s_{t+1} = S_j | s_t = S_i) = P(\overrightarrow{S_i\, S_j})$. It will later be useful to denote the set of transitions $\mathbf{E}$ and the size of this set as $|\mathbf{E}|$ or E.

In some cases (e.g. systems with real-valued variables) every observed state in the dataset will be unique, and the sequences for trials will not intersect in the state-space. Applying this methodology on such a representation will not produce any *useful*

[3]This will be true of the Markov representation even if the data is event-driven. The time interval is a resolution parameter of the Markov representation building process. Time-homogeneity can be relaxed with the appropriate modifications to the measures and algorithms.

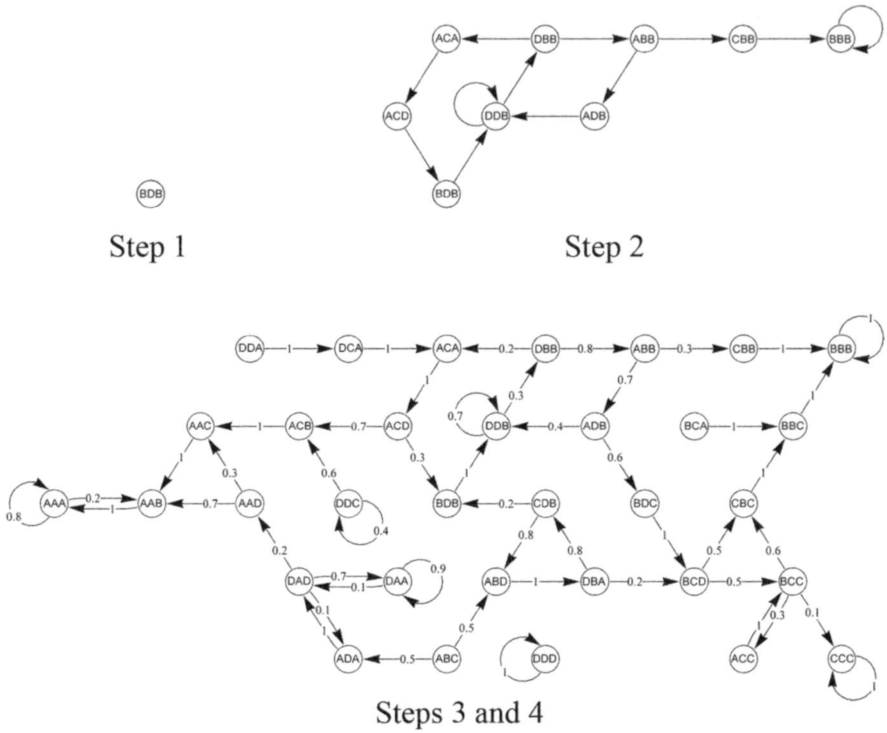

Fig. 1 The four steps in building a Markov model from data

results. In this case it is necessary to define value ranges for the variables such that all values within those bins are indistinguishable. Aside from reducing the number of configurations, this discretization lowers the resolution and information content of the Markov model description. However, by increasing the number of observations of each coarse-grained state and transition, binning improves the confidence levels of transition weights and our overall understanding of the aggregated patterns (see Sect. 7.3 for more discussion of these trade-offs).

Example 3.2 Instead of recording all four distinct actions as separate specifications, we could also/instead decide to record both $a1$ and $a2$ as $A1$, and $a3$ and $a4$ as $A2$. Then there are only $3^2 = 9$ possible states of the system. Formerly distinct configurations now overlap and observed state-changing transitions may now be self-edges. If, for example, both $a1$ and $a2$ are cooperating, and $a3$ and $a4$ are defecting actions, then this lower-resolution version of the system may provide more insight into cooperation/defection dynamics even though it does not distinguish the types of cooperation and defection.

4 Special States and Sets

To use a Markov diagram to represent system dynamics we will need to define various types of system behaviors in terms of system states, sets of states, and state transitions; and/or nodes, node sets, and edge weights. All features of graph elements and structure will be given probabilistic definitions in order for them to be compatible with the analysis approach used throughout. As a result, before the common features of system behavior can be incorporated, probabilistic definitions of the basic structural features they depend on must be first defined.

4.1 Paths

In graph theory a path (of length ℓ) is typically defined as a set of vertices and edges satisfying the schema $v_0, e_1, v_1, e_2, \ldots, v_{\ell-1}, e_\ell, v_\ell$ where the edge e_i links the vertex v_{i-1} to v_i [11]. Self-transitions, which represent both a lack of change and a change too small (i.e., within a bin) to count as a state change, are an important feature of Markov modeling and hence both nodes and edges may be repeated along paths. So 'path' as it is used here is the broader notion sometimes called a 'walk' in the graph theory literature. A path in a Markov model could be defined as a set satisfying the same schema used in graph theory, but we will use a slightly different definition to make the probabilistic aspects explicit.

Definition 3 A *path* is an ordered set of states such that from each state there exists a positive probability to transition to the successor state within the collection. A path from S_i to S_j is denoted $\tilde{\mathbf{S}}(S_i, S_j)$ or $\widetilde{S_i\ S_j}$.

To specify intermediate states (*markers*) for the system to pass through we can write $\widetilde{S_i\ S_k\ S_j}$ to denote a path from S_i to S_j that passes through (at least) S_k. To specify a long sequence of path markers we use the notation $\tilde{\mathbf{S}}(S_i, \ldots, S_k, \ldots, S_j)$.

To specify an exact path (i.e., a particular sequence of states) we adopt the notation $\overrightarrow{S_i\ \ldots\ S_j}$ for short sequences and $\overrightarrow{\mathbf{S}}(S_i, \ldots, S_k, S_{k'}, \ldots, S_j)$ for long ones. Thus a path from state S_i to state S_j going passing only through state S_k would be written $\overrightarrow{S_i\ S_k\ S_j}$ while one that stays in state S_i and S_k for two periods each would be $\overrightarrow{S_i\ S_i\ S_k\ S_k\ S_j}$.

Definition 4 The *length* of a path is the size of the set of transitions with positive probability among the set of ordered pairs of states along that path:

$$\ell(\overrightarrow{\mathbf{S}}(S_i, \ldots, S_j)) := |\{\overrightarrow{S_k\ S_{k'}} \in \mathbf{S} : P(\overrightarrow{S_k\ S_{k'}}) > 0\}|$$

This (possibly overly complicated) formal definition of length simply uses features of the definition of path above, but it is equivalent to the number of edges traversed along the path. There are reasons why a probabilistic formulation is preferred (even though superfluous in this case), but they are beyond the scope of the current presentation.

Observation 4.1 *It follows from the definition of a path that if there exists a path from S_i to S_j in the Markov model representation, then the probability is greater than zero of being in state S_j in the future given that the system is currently in S_i. Note that this feature may obtain even when no trial in the dataset contains a sequence connecting S_i and S_j; this possibility may be revealed through overlapping multiple sequences that intersect at intermediate states. Thus the possibility is an inference from the data based on the Markovian assumption.*

In the Markov construction, the probability of path $\overrightarrow{S_i\ S_k\ S_j}$ is the probability of making the two consecutive transitions $\overrightarrow{S_i\ S_k}$ and then $\overrightarrow{S_k\ S_j}$, which is just the product of those individual transition probabilities. Thus we can determine the probability of a path $\widetilde{S_i\ S_j}$ as

$$P(\widetilde{S_i\ S_j}) = \prod_{\overrightarrow{S_k\ S_{k'}} \in \widetilde{S_i\ S_j}} P(\overrightarrow{S_k\ S_{k'}})$$

Definition 5 A *cycle* is a path that starts and ends with the same state.

$$\widetilde{S_i\ S_i}$$

Corollary 4.1 *If a cycle has length one, then it represents a self-transition.*

$$\ell(\widetilde{S_i\ S_i}) = 1 \Rightarrow P(\overrightarrow{S_i\ S_i}) > 0$$

Addition graph structure measures such as diameter and girth can now also be given probabilistic definitions, but such work is left out of the current treatment. Although more structural features will be introduced as needed below, the definitions presented above suffice to move forward.

4.2 Landmarks in System Dynamics

This section provides definitions for common features of system dynamics in terms of the structural properties of Markov models that reveal them. Many of these features have existing definitions in terms of matrix operations or limiting distributions, but this method again uses probabilistic definitions based on the state transition representation.

Definition 6 A system state that always transitions to itself is called an *equilibrium* or a *fully stable state*.[4] An equilibrium e_i is a state S_i such that

[4]As is defined formally later, stability refers to a tendency to self-transition.

$$P(s_{t+1} = S_i | s_t = S_i) = 1.$$

In some cases a set of states (rather than a single state) plays a role similar to that of an equilibrium.

Definition 7 An *orbit* is a set of states such that if the system enters that set it will always revisit every member of the set and the system can never leave that set. Assuming $h > 0$ throughout **S** is an orbit if

$$\forall i \nexists h \; P(s_{t+h} = S_i \in \mathbf{S} | s_t \in \mathbf{S}) = 0$$

and

$$\forall i \; \forall h \; P(s_{t+h} = S_i \notin \mathbf{S} | s_t \in \mathbf{S}) = 0.$$

Theorem 4.2 *All the states in any orbit can be expressed as elements of an exact cycle; i.e., a cycle containing all and only elements of the orbit, though states may need to be repeated in that cycle identification.*

The proof of this theorem follows directly from the probabilistic definitions of a path, a cycle, and an orbit—specifically from the set of states for which there could be positive transition probabilities.

Definition 8 An *oscillator* is an orbit that is also a cycle with no states repeating along the path between the cycle's end states.

Definition 9 An *attractor* (denoted A_i) is either an equilibrium state or an orbit of the system.

In cases where attractor attainment or avoidance is the sole aim of using the model, we can collapse orbits into a single attractor state without loss of information. For this reason one can often use 'A_i' as if it were a single state, and that nomenclature is adopted here except in cases where being an orbit affects the analysis.

Though equilibrium analysis has a long and fruitful history, for many systems an equilibrium analysis is inappropriate. It is not the case that these systems fail to have attractors, or that they are difficult to identify, it's just that the goal of such systems is to remain in continual flux (e.g., homeostatic, dynamically stable, or autopoietic systems) and avoid so-called "point attractors" (i.e., attractors that incorporate a small percentage of the total number of states). The distinction between point attractors as just used and the formal definition of an attractor provided above is that point attractors identify "end points" for the system dynamics, whereas the definition here includes cases where the whole system is one giant attractor and the states are in continual change.

Because the Markov model represents all the observed behavior of the analyzed system, *every Markov model must have some set of states satisfying the conditions for being an attractor*. It is clear that there are no states outside the Markov model, and so some subset of **N** (possibly the whole set **N**) will form an orbit. A useful application of this methodology will therefore be to analyze the internal dynamics of an attractor set. One reason we may be interested in seeing orbits as the whole set instead of collapsing it into a single node is because within that orbit some states have a property we wish to keep track of. But more than that, discovering that a system's dynamics has only one attractor and every observed state is part of it provides significant insight into the behavior of that system. From that point forward, analyzing deeper details of the system requires a non-equilibrium analysis approach.

Definition 10 An attractor's *basin of attraction* is the set of states such that if the current system state is one of those states, then it will eventually move into that attractor. The basin of A_i or $\mathbf{B}(A_i)$ is a set of states $\mathbf{S}$ such that

$$\exists h \; P(s_{t+h} = A_i | s_t \in \mathbf{S}) = 1$$

Note that with this formal definition, the attractor is (correctly) a subset of its basin of attraction. However, often in discussion one is referring to the non-attractor part when referring to the basin of attraction. Some systems may spend a great deal of time in a basin of attraction before reaching the attractor located within it. This makes system behavior in the basin similar to an orbit itself, and one can get a deeper understanding of dynamics by determining the (probabilistic) time horizon after which each state in the basin is no longer visited. In such cases it is also sometimes helpful to utilize the following property to describe and make inferences about system behavior.

Observation 4.2 *Once in a basin of attraction the system can never leave it.*

$$\forall h \; P(s_{t+h} \in \mathbf{B} | s_t \in \mathbf{B}) = 1$$

Observation 4.3 *If a system only has one attractor, then every non-attractor state is in that attractor's basin of attraction.*

Observation 4.4 *If the systems has multiple attractors, then their basins of attraction are mutually exclusive.*

$$\exists j \; \forall k \;\; s_i \in \mathbf{B}_j \Rightarrow s_i \notin \mathbf{B}_k$$

Though basins of attraction are a common feature of analyzing system dynamics, a state need not be an attractor for the concept of a basin provided above to apply.

Definition 11 A state's *sink* is the set of states such that if the current system state is one of those states, then it will eventually move into that state. The sink of S_i is a set of states $\mathbf{S}$ such that

$$\exists h \; P(s_{t+h} = S_i | s_t \in \mathbf{S}) = 1$$

Attractors and basins are "absorbing" in the sense that they indicate sets of states that can be entered, but not left. A sink is a broader, non-absorbing notion along these lines. Being in the sink of S_i only implies that you can't leave until you visit S_i, but makes no restrictions on the system behavior after that. We can expand the definition of sink to apply to sets of states $\mathbf{S}'$ as well.

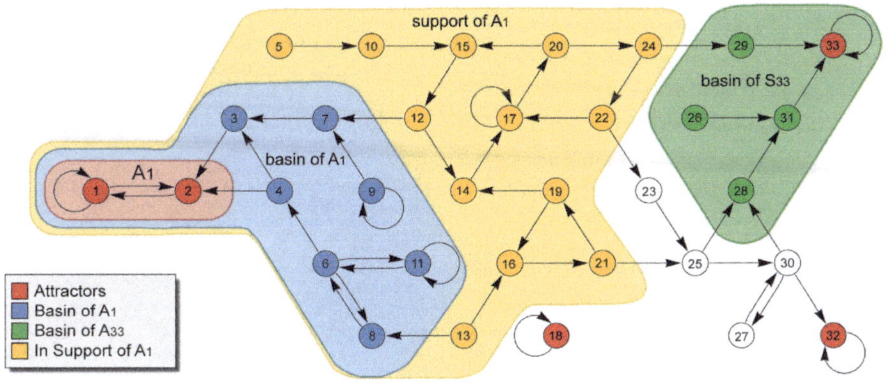

Fig. 2 The attractors of an example system, the basins of attraction for attractors A_1 and A_{33}, and the support of A_1

Definition 12 The *sink of a set* $\mathbf{S}'$ is the set of states that transition to some state within $\mathbf{S}'$ with certainty:

$$\{\mathbf{S} : \exists h \geq 0\ P(s_{t+h} \in \mathbf{S}' | s_t \in \mathbf{S}) = 1\}$$

Next we have an even broader notion of behavior tendency.

Definition 13 The *support* of a state (also known as its *in-component*) is the set of states which have a path to it. The support of S_i or $\mathbb{S}(S_i)$ can therefore also be thought of as the states with a positive probability of reaching S_i:

$$\mathbb{S}(S_i) = \{\mathbf{S}_j : P(\widetilde{S_j\ S_i}) > 0\}$$

And naturally we can define the support of sets of states as well.

Definition 14 The *support of a set* $\mathbf{S}'$, written $\mathbb{S}(\mathbf{S}')$, is the set of states that may transition into some element of $\mathbf{S}'$—which follows the previous definition for the existence of a path. We can therefore define it simply as the union of the supports of the elements of $\mathbf{S}'$.

$$\mathbb{S}(\mathbf{S}') = \{\mathbf{S} : \bigcup_{S_i \in \mathbf{S}'} \mathbb{S}(S_i)\}$$

Examples of the definitions of attractor, basin of attraction, and support are shown in Fig. 2.

If there are multiple attractors, then the *indeterminate states* of a system are the ones in multiple attractors' supports (i.e., ones that are not members of any basin of attraction). In this case, they capture a system's away-from-equilibrium dynamics and convey probabilistic information about the system's future possible states. However, recall that the whole system may be a single attractor and there may be no indeterminate states (but still non-equilibrium dynamics within the attractor).

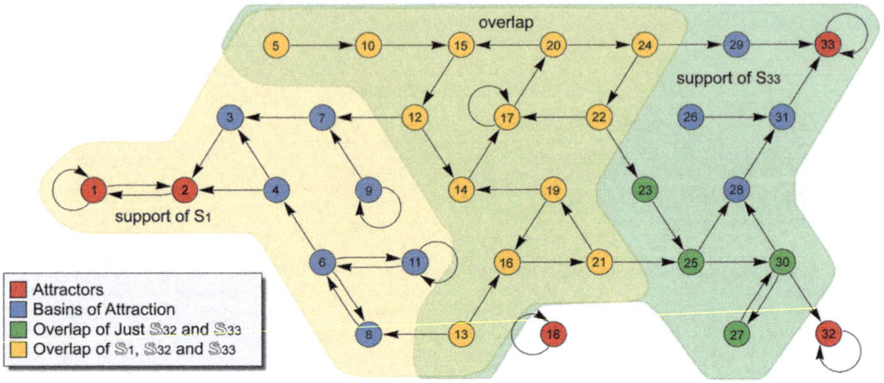

Fig. 3 The supports for the orbit containing S_1 and S_2 and the attractor S_{33} as well as their overlap

Definition 15 The *overlap* of a collection of states (whether attractors or not) is the set of states in all of their in-components; i.e., it is the intersection of the supports of the chosen sets. The overlap of $\{S_i, \ldots, S_j\}$, written $\Omega(S_i, \ldots, S_j)$, is the set of states in

$$\bigcap_{S_g \in \{S_i, \ldots, S_j\}} \mathbb{S}(S_g)\,.$$

An example showing the overlaps of two and three attractors can be seen in Fig. 3. We can use this definition of overlap to formally define the class of indeterminate states.

Definition 16 The *indeterminate states* of a system is the set of states in the union of the overlaps of all pairs of attractors (or whatever target states). That is, for each pair of states $(S_g, S_k) \in \{S_i, \ldots, S_j\}$ we have

$$\bigcup \Big(\bigcap \mathbb{S}(S_g)\mathbb{S}(S_k) \Big)$$

Observation 4.5 *The number of intermediate attractor states is always at least as large as the overlap of all attractors. For example, when $\mathbb{S}(S_i)$ and $\mathbb{S}(S_k)$ overlap, and $\mathbb{S}(S_k)$ and $\mathbb{S}(S_j)$ overlap, but $\mathbb{S}(S_i)$ and $\mathbb{S}(S_j)$ do not overlap, then $\Omega(S_i, S_k, S_j)$ is empty, but there are still indeterminate states in $\bigcup \Omega(S_i, S_k)\Omega(S_k, S_j)$.*

The overlap states are of interest because these are the states with positive probabilities for ending up in each of the states of interest (usually attractors). The overlap states for any subset of attractors are therefore among the indeterminate states of the system. Any state not in the overlap for some set of attractors is in some attractor's basin of attraction. So even though the overlap of attractors is merely a structural feature of the generated transition diagram (without even considering the weights) it identifies the set(s) of states that must be maintained in order to avoid an equilibrium outcome. The maintenance of the system in states within the overlap of attractors is therefore already a simple notion of robustness or sustainability, and we will revisit this notion as energy levels in Definition 37.

Observation 4.6 *If no attractors' supports overlap, then every attractor's support is just its basin of attraction and the system has a deterministic outcome given any initial state.*

Understanding that a system has a deterministic outcome does not render the mapping of the dynamics leading up to that outcome uninformative. Of course it may lead one to redesign their model, but system behavior within a basin of attraction and/or attractor can be a complicated mix of dynamics. Some property of interest may be gained and lost at different states within the basin and the churning of the aspect (variable) values for long durations may be key to understanding some phenomena.

We now continue with structural features, but now we're going to look at exit dynamics instead in entrance dynamics.

Definition 17 A state's *out-degree* is the number of distinct successor states (states that may be immediately transitioned into). The out-degree k_i of state S_i equals

$$|\{S_j : P(s_{t+1} = S_j | s_t = S_i) > 0\}|$$

At times we will find it useful to use S_k to denote a neighboring state and $\mathbf{S_k}$ to denote the set of neighboring states. Nothing in our analysis depends on the in-degree, so we do not have any special notation for that.

Definition 18 The *reach* of a state (also called its *out-component*) is the set of states that the system may enter by following some sequence of transitions; i.e., *all possible* future states given an initial state. The reach of S_i or $\mathbf{R}(S_i)$ is the set

$$\{S_j : \exists h \; P(s_{t+h} = S_j | s_t = S_i) > 0\}$$

Theorem 4.3 *The number of states in every successor state's reach is less than or equal to the number in the initial state's reach.*

$$\forall i, j \; \widetilde{S_i \, S_j} \Rightarrow |\mathbf{R}(S_i)| \geq |\mathbf{R}(S_j)|$$

Theorem 4.3 generalizes to all paths (which is just a sequence of transitions) so that reach never increases as the systems transitions along any path. The proof follows from the facts that (1) any state reachable from S_j is reachable from S_i by first transitioning into S_j, and (2) there may be an S_h such that $\overrightarrow{S_i \, S_h}$ and there is no path from S_j to S_h. Furthermore, $|\mathbf{R}(S_i)| = |\mathbf{R}(S_j)|$ only when there exists paths $\widetilde{S_i \, S_j}$ and $\widetilde{S_j \, S_i}$.

A *strongly connected component* of a directed network is a set of nodes such that there is a path from every node in the set to every node in the set (including itself).

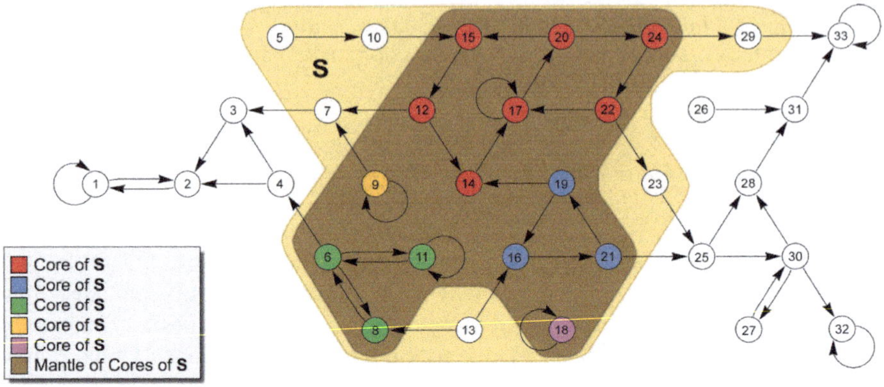

Fig. 4 Assume the set of states **S** possess a property of interest. Within that set one can identify five cores: one is an equilibrium, another is a single state, one is a simple cycle, and the other two have more complicated dynamics

We will find the same construct useful, but this paper adopts a different name and a probabilistic definition.

Definition 19 A *core* of a set is a maximal subset wherein every member of the subset is in the reach of every member of the subset. A core of some set **S** is written $\mathbb{C}_a(\mathbf{S})$ and is a subset satisfying the condition:

$$\bigcap_{S_i \in \mathbf{S}} \bigcap_{S_j \in \mathbf{S}} S_j \in \mathbf{R}(S_i)$$

Some sets will have multiple cores and they are indexed by a—the set of **S**'s cores can be called **S**'s *mantle*. An example of the the cores making up a mantle can be seen in Fig. 4.

Observation 4.7 *Every state in a core has the same reach.*

Another structural definition that we will find immediately and intuitively useful identifies boundary states for system properties in the following way:

Definition 20 The *perimeter* of a set, $\mathbb{P}(\mathbf{S})$, is a collection of those states in the set that may transition to states outside the set. That is, $\mathbb{P}(\mathbf{S})$ is the set of states such that

$$P(s_{t+1} \notin \mathbf{S} | s_t \in \mathbf{S}) > 0$$

In keeping with the core and mantle analogy, the perimeter states of the mantle of **S** could be referred to as **S**'s *crust* and can be seen in Fig. 5. Though a purely structural feature, the perimeter states themselves, without further specification, describe one commonly deployed concept of tipping points.

In the next section we apply the above definitions in different combinations and different contexts to identify system behaviors related to tipping and thresholds. We

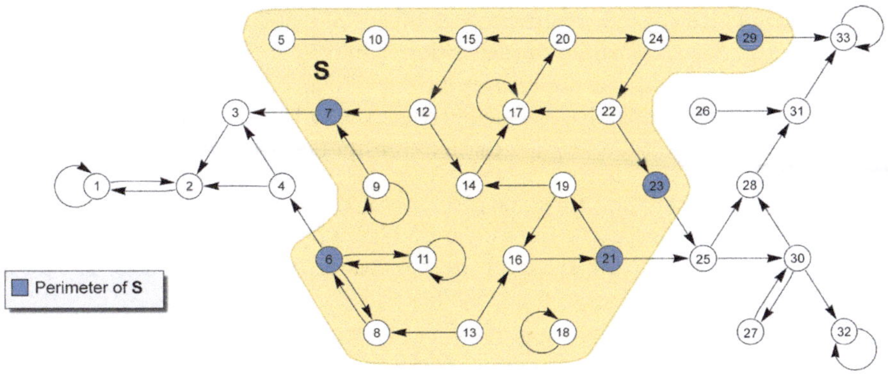

Fig. 5 The perimeter states of the chosen set of states **S**

will soon see how specifying different base sets produces different types of tips, and also how to incorporate probabilities into our characterization of tipping behavior.

5 Tipping Phenomena and Related Concepts

Using Markov models and the special states and sets defined above as a springboard, this section defines and briefly describes several terms related to the concept (or more to the point, concepts) of tipping points. Critical phenomena and tipping points of various kinds share the defining feature that (for whatever reason) a characteristic of the system is different before and after some change. The tipped characteristic may be a property such as system performance or an aggregate measure, or it may be the behavior itself. There are, of course, many ways in which system and dynamical properties can differ, and each way is a different kind of tipping phenomenon.

5.1 Levers

The analysis of tipping phenomena starts with individual state aspects. Recall that state changes occur if and only if there is a change in some aspect of the current state.

Definition 21 The *levers* of a state are the aspects of that state such that a change in those aspects coincides with a change in the system's state. The levers of S_i, denoted $L(S_i)$, is the set of aspects in

$$\bigcup_{j=1}^{k} \bigcup_{h=1}^{Q} X_{h(i)} \neq X_{h(j)}.$$

The levers of a given a state S_i are those aspects of S_i that are different in any of the k states that S_i has a transition to (network neighbors) in the Markov model, thus each state has k distinct levers. For cases in which neighboring states differ by more than one aspect, the respective element of the set $L(S_i)$ is the whole set of aspects that need to change as a single lever for that transition.

Definition 22 A *lever point* for some aspect (or set of aspects) is a particular transition for which that aspect is a lever. An aspect's set of *lever points* is the collection of transitions that a change in that aspect (or those aspects) alone generates. The lever points of aspect X_h is the set of transition in

$$\bigcup_{\overrightarrow{S_i\ S_j}}^{\mathbf{E}} X_{h(i)} \neq X_{h(j)}.$$

Levers and lever points work complementarily: for levers we pick a state and find the aspects that change across its outgoing transitions, and for lever points we pick an aspect and find all the transitions corresponding to changes in that aspect. These definitions can also be bounded to collect more specific information such as finding all the transitions corresponding to X_h changing from x_1 to x_2 rather than just any change in X_h (see Definition 26 for more detail on this refinement).

It is occasionally helpful to refer to the aspect change(s) that generate a specific transition.

Definition 23 $L(\overrightarrow{S_i\ S_j})$ symbolizes the *lever set* of $\overrightarrow{S_i\ S_j}$: the aspect or aspects that differ between S_i and S_j.

In some applications we will be interested in how many aspects change for a transition.

Definition 24 The *magnitude* of the lever set of a specific transitions is $|L(\overrightarrow{S_i\ S_j})|$.

Though levers as they are defined here do not depend on the ability to control that aspect, the choice of 'lever' for this concept is motivated by the realization that in some models control of some aspects is available. One may be performing a tipping point analysis precisely because one is choosing levers to bring about one state versus another (or agents within the model may be choosing).

Example 5.1 Imagine a model wherein each aspect is a variable representing some part of a policy (e.g., amount of money spent on each line item). Each aspect change has an associated cost (legal, bureaucratic, time, etc.). The modeler may be trying to determine the lowest cost, feasible route from the current policy to some desired policy; or perhaps to determine how far policy can be changed on a specific budget. The cumulative magnitudes of lever sets along a path may adequately approximate such a cost measure. In general the sum of the magnitudes along a path is a rough measure of how difficult it is for the system to behave that way. Techniques from circuit design applied to the Markov model may be gainfully applied to such models.

In some contexts we may wish to know how much change an aspect is responsible for across the system's dynamics.

Definition 25 The *lever strength* of an aspect is the normalized sum of the probabilities of all transitions for which it is a lever. The lever strength of X_h is equal to

$$\frac{1}{N}\sum_{\overrightarrow{S_i\,S_j}}^{\mathbf{E}} P(\overrightarrow{S_i\,S_j}|X_{h(i)} \neq X_{h(j)}).$$

Note that since the exit probabilities for each node must sum to one, dividing by the number of nodes N normalizes the lever strength measure to the total probability density in the model to produce a scale-free measure.

Observation 5.1 *It is therefore also the case that the sum over lever strengths for all levers equals* N.

The lever strength measure could be used to determine which aspects to control in order to maximize (or minimize) one's ability to manipulate the system's behavior. It may also be related to the cost of letting that aspect vary over time since it measures the behavioral impact of that aspect. For all the lever-related phenomena, narrowing the domain of states from the whole system to a set possessing a property of interest focuses the analysis in the obvious way to transitions within the chosen set.

5.2 *Thresholds*

In some cases we are interested in more than just which aspects change across a transition, but also in the specific changes in **values** of levers across transitions.[5]

Definition 26 A specific value for an aspect is a *threshold* if the value of the aspect changes from that value across a transition. The value x is a threshold of X_h if

$$\exists \overrightarrow{S_i\,S_j}\; X_{h(i)} = x \text{ and } X_{h(j)} \neq x.$$

This definition can be applied *mutatis mutandis* for a set of values (e.g. for a lever set) which can be distinguished by the name *threshold line* when appropriate. In applications one may want to restrict the domain to (for example) distinguish threshold values for transitions inside and outside some set of states.

If there are multiple transitions differing at the same threshold value, then uncovering this feature refines our understanding of that lever's role in system dynamics. Thus determining the threshold value for one transition is merely a means to the end of determining the strength of the levers with that threshold.

---

[5]Recall that many of the things that can be included as aspects of states are not numeric parameters, and so what counts as a "value" for that aspect is meant to be interpreted broadly.

Definition 27 The *threshold strength* of x is the normalized strength of the levers for which x is the threshold value:

$$\frac{1}{N}\sum_{\overrightarrow{S_i\,S_j}}^{\mathbf{E}} P(\overrightarrow{S_i\,S_j}|X_{h(i)} = x \text{ and } X_{h(j)} \neq x.)$$

If a particular value for a particular aspect plays a large role in system dynamics then crossing that threshold is another oft-used version of tipping point behavior.

These definitions of threshold and threshold strength only require that the end state's value be different from the start state's value. In common usage, however, thresholds establish different and separate boundary values for ascending and descending values. If a threshold only affects system dynamics in one direction then we can determine that from the Markov model using the following definitions.

Definition 28 An *upper bound threshold* of X_h is a value x such that

$$\exists\overrightarrow{S_i\,S_j}\; X_{h(i)} \leq x \text{ and } X_{h(j)} > x.$$

Definition 29 A *lower bound threshold* of X_h is a value x such that

$$\exists\overrightarrow{S_i\,S_j}\; X_{h(i)} \geq x \text{ and } X_{h(j)} < x.$$

The threshold strength measure can be adapted to these ascending and descending definitions in the obvious ways. Sets satisfying these definitions can tell us how frequently crossing that threshold in that direction acts as a lever. For example, if one uncovers that all transitions involving X_h as a lever do so with an increase in X_h from below some x to above that x, then that becomes an indicator that the role of x deserves further investigation. Upper and lower bound thresholds satisfy another common usage of "tipping point".

Example 5.2 This general definition admits examples from many different kinds of systems and can even apply to parts of systems (such as agents). In Granovetter's model of riot spreading [8] we can talk of each agent having its own threshold—the number of rioting agents necessary to make each agent join the riot. This is just the same threshold definition applied to a lever set where the levers happen to be the same feature of each agent. In Granovetter's model the threshold value is the same in both directions.

We can also talk of thresholds in the properties of the system dynamics that track how the system transitions through states. Instead of being a value for an aspect within the model, it would be a value for one of the measures defined for properties of system dynamics.

5.3 Critical Behavior

In some system analyses the property of interest is, in the most general sense, what is available for the future. If one does not know much about a system's dynamics then even knowing how many states could potentially be transitioned into (i.e., the size of the reach) provides an informational benefit. The measures below become increasingly refined and detailed, but we start with some simple measures.

Definition 30 A state's *stretch* is the number of states in its reach. So the stretch of S_i equals

$$|\mathbf{R}(S_i)|.$$

Definition 31 A system dynamic (i.e., a particular state transition or path) is considered *critical behavior* if and only if it produces a decrease in stretch; that is, critical behavior is any $\overrightarrow{S_i\, S_j}$ such that

$$|\mathbf{R}(S_i)| > |\mathbf{R}(S_j)|.$$

Recalling Theorem 4.3 that reach can not increase along any path, critical behaviors are those that limit the system's future states. We can also measure *how critical* each transition is. Subtracting the end state's stretch from the start state's stretch provides such a measure, but it is not scale-free (because both the range of values and the particular value for this measure depends on the total size of the system) and so cannot be readily compared across different systems. We can normalize the stretch difference with the size of the system to which it is being applied to produce a percentage measure.

Definition 32 The *stretch-gap* of a transition is the change in the percent of the total number of states that can be reached. This quantity equals

$$\frac{|\mathbf{R}(S_i)|}{N} - \frac{|\mathbf{R}(S_j)|}{N}.$$

Because this measure includes the total number of states in the system it clearly is not scale-free either. However despite this limitation it does provide information about the system's future and is an intuitive way to compare transitions within the same system—even at different resolutions. As Example 5.3 below demonstrates the stretch-gap reports how much of the system's state space is cut off by each transition and this information could be used, for example, to manipulate system dynamics to prolong system longevity. We also have an alternative, fully scale-free, measure of the reduction in reach across a transition.

Definition 33 A **transition's** *criticality* is one minus the ratio of the start and end states' stretch. The criticality of $\overrightarrow{S_i\, S_j}$ equals

$$1 - \frac{|\mathbf{R}(S_j)|}{|\mathbf{R}(S_i)|}.$$

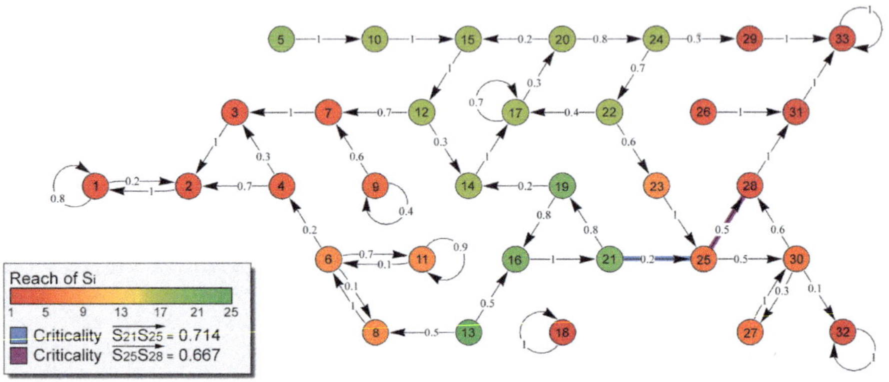

Fig. 6 Color-coded reach values allows one to visualize the Stretch-Gap for each transition. Criticality measurements are provided for two of the transitions reporting the percent of future states that become unavailable with those transitions

An example of two critical transitions appears in Fig. 6. Again from Theorem 4.3 we have the following corollary regarding the range of values for the ratio of reaches.

Corollary 5.1 *A transition's criticality will be between zero and one.*

Proof. Let $a = |\mathbf{R}(S_j)|$ and $b = |\mathbf{R}(S_i)|$ for transition $\overrightarrow{S_i\,S_j}$. From Theorem 4.3, $a \leq b$. If $a = b$, then $\frac{a}{b} = 1$ which yields a criticality of zero. For $a < b$ we can decrease a or increase b to find the other bound, but since a and b are natural numbers increasing b is the better approach. Using a well-known mathematical fact suffices for finding the other bound: $\forall a \ \lim_{b\to\infty} \frac{a}{b} = 0$.

Observation 5.2 *Transitions within a cycle (which includes self-transitions) always have zero stretch-gap and zero criticality.*

The measure of criticality just provided therefore coheres with the concept of criticality insofar as any transition that has no affect on the set of states that may be visited in the future is not be a critical transition. A transition is critical if and only if the possible system dynamics are irreversibly changed across it. Critical transitions are another oft-used tipping concept; however, criticality (as used here) measures changes in the dynamics themselves rather than changes in the dynamics of some system property (which is covered in Sect. 5.4).

Example 5.3 The system represented in Fig. 6 has thirty-three states in total, each one color-coded by its stretch. Compare the pattern in node shading to the set of states comprising the attractors, basins, and support in Fig. 2 and the overlap in Fig. 3. Stretch alone, though a simplistic structural measure, is often effective in partitioning the system dynamics into regions of similar behavior. Stretch-gap performs well as a discriminator that groups states into these regions because for systems with sufficiently sized basins, transitions into one precludes all the states in all other basins.

To wit, stretch drops between zero and four states within any basin or overlap, but drops four to eighteen states crossing a boundary (in this example). This effectiveness obviously depends on the structure of the Markov model, but for many systems this simple technique may provide all the information required; and it may be the best one can do with available data (in term of confidence over the measures). The reason we may want to measure stretch and criticality is because in addition to detecting basins and different overlaps, it also reveals progression through phases of behavior marked by a series of cores (see Example 6.4 for more detail).

S_{21} has a stretch of 21, S_{25} has a stretch of 6, and S_{28} has a stretch of 2. There are 33 states in this system so the stretch-gaps of $\overrightarrow{S_{21}\ S_{25}}$ and $\overrightarrow{S_{25}\ S_{28}}$ are 45.45% and 12.12% respectively. That means that 45.45% fewer of the system's states can be reached after the $\overrightarrow{S_{25}\ S_{28}}$ transition. We can also use this to determine the stretch-gap of $\widetilde{S_{21}\ S_{28}}$ as 57.58% regardless of the particular path taken. Only the start and end states' stretches are necessary to calculate this, but the result is always equal to the sum of the stretch gaps of each transition taken.

Let's compare these figures to the criticality of the same transitions. The criticality of $\overrightarrow{S_{21}\ S_{25}}$ is 0.714 and the criticality of $\overrightarrow{S_{25}\ S_{28}}$ is 0.667. That means that the system only has 71.5% of the possible future states in state S_{25} as it did in S_{21}. A composite measure is also possible for the criticality of $\widetilde{S_{21}\ S_{28}}$. It can be calculated just using the start and end states' stretches, using the standard percentage of a percentage of a percentage … calculation.[6] So the criticality of $\widetilde{S_{21}\ S_{28}} = 0.714 + 0.667 \cdot (1 - 0.714) = 0.905$.

These measures above are intended to be just rough measures useful in certain limited contexts and when information about the system is limited. For starters, these measures consider only the structure of the Markov models, not the probabilities. Also, they apply to transitions rather than states. Both limitations are now overcome with richer definitions of the relevant dynamical properties.

Definition 34 The *criticality* of a **state** is the probabilistically weighted sum of the criticality of all the transitions from that state. So to find the criticality of S_i we calculate

$$\sum_{j=1}^{k} P_{ij}\left(1 - \frac{|\mathbf{R}(S_j)|}{|\mathbf{R}(S_i)|}\right)$$

where by convention of the use of k that is the sum over S_i's neighbors.[7]

Because by definition of a Markov model the sum of the exit probabilities sum to one, state criticality will also be a scale-free measure with values between zero and

[6] The value equals the iterated sum of the previous transition's criticality and the product of the transition criticality with the previous transition's criticality's complement.

[7] In this case it does not matter whether the sum is limited to span over neighbors or all the vertices because $P_{ij} = 0$ for S_j that are not neighbors. This convention will be used throughout—including cases where limiting an operation to neighbors matters.

one. All these criticality measures quantify the constriction of future possibilities on a state-by-state basis which is useful if we want "to keep our options open". Sometimes that is exactly what we want to measure, but sometimes we will want to measure system dynamics with reference to some particular features and that is what the following definitions for tipping points allow us to measure.

5.3.1 Critical Levers

As a refinement of levers from Definition 21, we can apply the lever concept to critical states to identify the relationship of a particular aspect to a system's dynamics.

Definition 35 A state has a *critical lever* if a change in that aspect (or those aspects) of the state will reduce the reach.

This merely combines the concept of a lever with the concept of critical behavior (Definition 31). By looking more deeply at the aspects driving the state changes and calculating the magnitude and strength of different critical levers we can gain a better understanding of how microfeatures generate the macrobehavior of the model.

5.4 *Tipping Behavior*

As mentioned in the Introduction, the common feature of the measures in this section is that some states or transitions mark a shift in the properties of a system or of its states. For the criticality measures above the difference was the number of reachable states. The following measures generalize to any sets distinguished by a chosen characteristic. Given states exhaustively compartmentalized (i.e., partitioned) by the property (or properties) of interest the following techniques can find where shifts occur and measure their magnitude.

5.4.1 Tipping Points

For some models we are interested in the achievement of a particular state (e.g., an equilibrium or achieving some goal in terms of a particular value for some aspect) or a particular system behavior (e.g., a path linking two states). We denote the particular state (or set) of interest as the *reference state* (or *reference set*). Below we will see examples of specific reference states (e.g., attractors and functional states) but first the general case. There are many ways in which behavior may change with respect to a reference state or set (e.g., probability of reaching it, probability of returning to it, or probability of visiting an intermediate state): each property may partition the states into different equivalence classes (groups with the same value of the property). *It is the movement between equivalence classes that counts as tipping behavior.*

Definition 36 A *tipping point* is a state which is in the perimeter of an equivalence class for some property.

Any selected property value or value range can identify a set of states as a reference set; i.e., the subset of state having the property value(s) of interest. Recall from Definition 20 that perimeter states are those states within a specified set from which the system's dynamics can leave that specified set. Because the sets here are determined by the property values, leaving a set implies a change in that property—and that is the basic characterization of a tip. This definition does not eliminate the possibility that the system could tip back; i.e. a property value tipped out of may later return. That possibility depends on what property is establishing the equivalence classes and the structure of the system dynamics.

Example 5.4 Consider a climate change model that includes variables for many features of the environment and relates the CO_2 content of the atmosphere to global temperature, sea level, precipitation, glacial coverage, etc. Due to feedback mechanisms in the system, values for environmental variables change through "punctuated equilibria",[8] and thus produce equivalence classes for each property. *Ex hypothesi* people can manipulate the level of CO_2 to higher or lower values. The upper and lower thresholds for CO_2 values may differ for many of these environmental properties (which may be something like a glacial coverage of between 100 and 200 km^2), but the point is that CO_2 levels could raise temperatures to the point where glaciers disappear and then later lower past the point where glaciers will form again.

This example demonstrates that system behavior can tip out of an equivalence class and then later tip back in—in contradistinction to how the term is sometimes used. Phase transitions in condensed matter physics are another example of reversible tipping behavior. So while some have posited that tipping points are points of no return for system behavior, that turns out to be true only for certain systems and is not properly part of the definition of tipping behavior. One can, however, combine the definitions of tipping points and sinks to produce a refined concept of a one-way tip in an obvious way.

5.4.2 Energy Levels

Dynamics of staying, leaving, returning, and avoiding a specified set of states are covered in Sect. 6. Here we continue with ways to quantify changes in what is possible for system dynamics for different states and transitions. These measures apply for any reference state or reference set, but for convenience and to foster building intuitions the following presentation will adopt the notation of attractor (A_i) for a reference state or reference set. **A** is used for a collection of reference objects; i.e., a set of

[8]The phrase "punctuated equilibrium" is in quotes here because I will use it in a more formal sense in Sect. 6.4, and though the variable changes also have the feature of being stable for long periods of time between abrupt changes, the internal dynamics within an equivalence class for these properties probably do not satisfy equilibrium-like conditions.

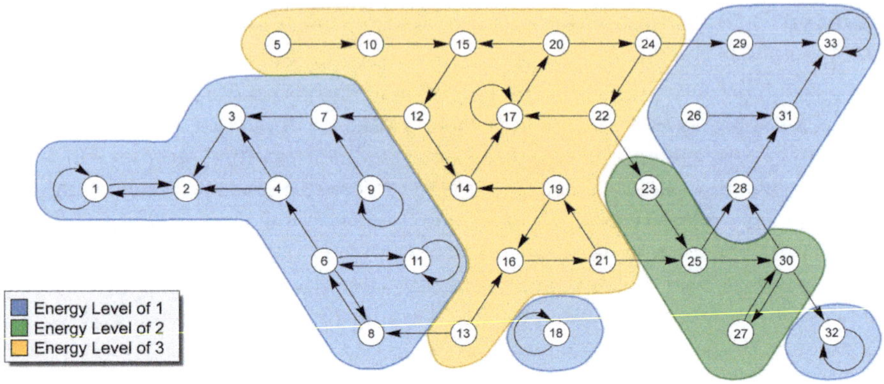

Fig. 7 The energy plateaus are the equivalence classes formed by using attractors/orbits as reference states

distinct A_is. Using the example diagram in Fig. 2, we can identify $A_1 = \{S_1, S_2\}$, $A_2 = S_{18}$, $A_3 = S_{32}$, and $A_4 = S_{33}$, and then using the collection of attractors as our reference set, **A** is $\{A_1, A_2, A_3, A_4\}$.

Definition 37 The *energy level* of a state is the number of reference states within its reach. We write this as $E(S_i)$ and it equals

$$|\bigcup_{A_i}^{\mathbf{A}} A_i \in \mathbf{R}(S_i)|$$

Energy level quantities partition the system's states into equivalence classes.

Definition 38 The equivalence class mapping created by states' energy levels is called the system's *energy plateaus*. Each energy plateau is a set

$$\bigcup_{i,j}^{N} E(S_i) = E(S_j).$$

The energy plateaus of our running example Markov diagram appears in Fig. 7.

Definition 39 The change in energy across a transition is called an *energy precipice* or *energy drop*. We can measure the magnitude of an energy precipice in the obvious way:

$$\triangle E(\overrightarrow{S_i\ S_j}) = E(S_i) - E(S_j)$$

Theorem 5.2 *An energy precipice is never negative:* $\triangle E(\overrightarrow{S_i\ S_j}) \geq 0$.

Proof. From Definition 37 the energy level of any state S_i is the number of attractor sets in $\mathbf{R}(S_i)$. By Theorem 4.3 for any transition $\overrightarrow{S_i\ S_j}$ it is always the case that $\mathbf{R}(S_j) \subseteq \mathbf{R}(S_i)$. This decreasing reach property implies

$$|\bigcup_{A_i}^{\mathbf{A}} A_i \in \mathbf{R}(S_j)| \leq |\bigcup_{A_i}^{\mathbf{A}} A_i \in \mathbf{R}(S_i)| \Rightarrow E(S_j) \leq E(S_i).$$

Since the end state of a transition always has a lower or equal energy level, $E(S_i) - E(S_j)$ is always greater than or equal to zero.

The energy drop is equal to zero when the transition is within an energy plateau.

We can now measure the degree to which a state is likely to be the site of a tip, starting with a tip out of an energy plateau.

Definition 40 The *energy tippiness* of a state is the probabilistically weighted proportional drops in energy across its immediate successors:

$$1 - \sum_{j=1}^{k} P_{ij} \left(\frac{E(S_j)}{E(S_i)} \right)$$

Different reference sets will obviously alter the energy level equivalence classes, and hence the tippiness measures of states. One can, for example, use the states with $X_h \geq x$ as the reference states. Then a drop in energy refers to having fewer states with that property as potential future states. The energy tippiness measure therefore reveals which states the maintenance of that property is most sensitive to, *and how sensitive the property is at each state.*

Theorem 5.3 *Energy tippiness ranges from zero to one.*

Proof. The lower bound occurs when all neighbors can reach the same number of reference states: $\forall k\ E(S_i) = E(S_k)$. In this case tippiness equals $1 - \sum_{j=1}^{k} P_{ij} \cdot 1 = 0$ since exit probabilities must sum to one. By Theorem 5.2 we know $E(S_i) \not< E(S_k)$. When $E(S_i) > E(S_k)$ the upper bound occurs when a state has every reference object and only reference objects as neighbors. We know from the definition of a Markov model that $\sum_{j=1}^{k} P_{ik} = 1$. Reference objects have an energy level of one and the energy level of S_i in this case is $|\mathbf{A}|$ so S_i's tippiness is $1 - \sum_{j=1}^{k} \frac{P_{ij}}{|\mathbf{A}|} = 1 - \frac{1}{|\mathbf{A}|}$. Thus the upper bound of S_i's tippiness goes to 1 as $|\mathbf{A}| \to \infty$.

Note that tippiness uses the ratio of energies rather than the difference; this makes tippiness a dimensionless metric and thus comparable across any state or system. Sometimes one will be more interested in minimizing the magnitude of energy drops, or avoiding states with the highest expected magnitude of energy drops, which have obvious formulations given the above definitions.

5.4.3 Tippiness

Recall that tipping points are potential exits from sets identified as equivalence classes for some property. While energy tippiness measures the sensitivity to the number of states possessing some identified property, we may also wish to measure sensitivity of the property itself.

Definition 41 The *tippiness* of a state S_i with respect to property X_h is the probabilistically weighted changes in the aspect's value across its immediate successors:

$$\sum_{j=1}^{k} P_{ij} |X_{h(i)} - X_{h(j)}|$$

We can further refine this measure in many ways depending on the intended use. For example, one may restrict the calculation to neighbor states with $X_{h(j)} < x$ to determine the sensitivity at S_i of that aspect dropping below the value x. One may also combine aspects in some appropriate way (e.g., multiplying two aspects) or measure some complicated system property and apply tippiness to that instead of a value of the system itself. This measure is not scale-free because it depends on the magnitudes of the aspect values, but it is useful for evaluating and comparing states within a system. If the aspect value is properly normalized, then a measure similar to energy tippiness can be used to produce a scale-free measure.

5.4.4 Tipping Levers

As another refinement of levers from Definition 21, we can apply the lever concept to tipping points to relate tips to the individual aspect changes that drive them.

Definition 42 X_h is a *tipping lever* of some specified set of states **S** if a change in X_h transitions the system out of **S**.

Thus a tipping lever is a combination of the definition of a lever with the concept of tipping behavior. While all states except equilibria have ordinary levers, only perimeter states for some set have tipping levers. Identifying the variables and values that must change to shift system properties is precisely what we'd like a "tipping point" analysis to reveal. These ideas are further refined in the analysis of robustness that follows.

6 Robustness-Related Measures

This section will use the Markov model framework provided above to establish formal definitions of several related concepts: robust, reliable, sustainable, resilient, recoverable, stable, and static; as well as their counterparts: susceptible, vulnerable, and fragile. Though it is unlikely that any mathematically precise definition will maintain all the nuances of the full concept sharing the same name, I have tried to pick terms that match as best as possible the dynamic being defined. The important thing is that the formulation here are useful, we can easily refer to them using these terms, and we can *reach a common understanding of the dynamics being referred to.*

6.1 Stable, Static, and Turbulent

The measures in this first group are conceptually simple with intuitive mathematical definitions and straightforward algorithms. They nevertheless identify important features of system dynamics and act as building blocks for more sophisticated measures.

Definition 43 A **state's** *stability* is how likely that state is to self-transition. S_i's stability is

$$P(s_{t+1} = S_i | s_t = S_i).$$

While this may seem a trivial property, it is consistent with a useful distinction from system dynamics: the difference between stable and unstable equilibria. Because the Markov models used here are finite and discrete, the binning of values used to generate the model's states can produce an attractor state that includes a neighborhood of aspect values around some equilibrium point's values. Thus the system behavior given that state includes the system's response to small perturbations to (or variations around) the equilibrium point values. States containing a stable equilibrium will tend to stay within this neighborhood because values close to the equilibrium tend to move toward the equilibrium, and this is reflected in a self-transition probability close to or equaling one. Since values that are nearby an unstable equilibrium, but not exactly on the equilibrium point values, will tend to move away from the equilibrium values, we would see this reflected in comparatively low self-transition probabilities (i.e., the node in the Markov model would not actually be an attractor even if a change in the binning of values would produce an attractor with similar aspect values). These results exactly match attributions of stability and instability in the Markov model via this definition of state stability.

We can extend the concept of stability to sets of states in two ways: (1) transitions that fail to leave that set and (2) sets containing stable states. We start with the first of these two.

Definition 44 The *stability* of a **set** is the probability that the system will not transition out of the set given that the system starts within the set. We calculate this as the average of the compliment to the individual states' exit probabilities. So set stability is

$$\frac{1}{|\mathbf{S}|} \sum_{S_i \in \mathbf{S}} 1 - P(s_{t+1} \notin \mathbf{S} | s_t = S_i).$$

This is a crude measure because is doesn't properly reflect the probability of staying within the set over time, it only looks one time period ahead. A more sophisticated notion of staying within a set of states (called 'sustainability') is presented by Definition 50 below. In some cases this measure suffices to understand the maintenance of a property; furthermore, because it requires less data to achieve acceptable confidence levels and has a much lower algorithmic complexity (and hence computational time), this measure is preferred for some applications.

The word 'static' is distinct from, but related to, stability and is often used to indicate a lack of dynamics in a system. That is the sense attached to the following formal definition extending the measure of state stability to the content of sets. Recalling that self-transitions are a lack of state change for any reason (delay, autopoiesis, a change smaller than the state resolution, rigidity, etc.) staticness aggregates the lack of state-changing transitions within a set.

Definition 45 The degree to which a set is *static* is the average of the states' stability values:

$$\frac{1}{|\mathbf{S}|}\sum_{S_i \in \mathbf{S}} P(s_{t+1} = S_i | s_t = S_i)$$

This definition, though simple, captures how likely a system can be expected to be in the same set of states for consecutive time steps in a way that is comparable across sets and across systems with different numbers of states. What this definition fails to capture is that sets containing equilibria will spend an infinite amount of time in the equilibrium states, whereas sets lacking equilibria will have upper bounds on how much time is spent in the set. And yet, because this measure uses average state stability, it is easy to construct cases where an equilibrating systems has a lower static level on the given definition. This is not a flaw in the measure as much as it is a caveat for when to apply it.

Static and stable set measurements are similar in their calculation but distinct in their sense. Set stability is a measurement of lack of change with respect to being in that set—it is a lack of change out of a set (though it ignores dynamics that stay within the set). It is therefore only applicable if the set chosen is smaller than the whole system. Staticness only measures dynamics internal to the set (in fact internal to each state in that set), and so it can apply to the whole system. Staticness may be more useful for comparing systems' overall level of system *dynamism* (related to *turbulence* discussed in Sect. 6.1.1) than it is for measuring the maintenance of a property (i.e., the shared features of the states within the chosen set).

Theorem 6.1 *If* $\mathbf{S} = \{S_1\}$ *(i.e., is a set with one state) then the static measurement equals the stability measurement.*

This theorem clearly follows from the fact that for a set of one state the sum of transitions staying within the set equals the sum of self-transitions (which is just the weight of the self-transition). Also in this case set stability naturally equals state stability because the state is the whole set.

For sets larger than a single state, however, set staticness bounds set stability in some structures, and sometimes the other way around. One can easily imagine the case of two states forming an oscillator such that set stability is one and set staticness is zero. However, if set stability is less than one, then staticness cannot be one because at least some perimeter states are not equilibria. Conversely, if some perimeter states have staticness less than one because they have transitions outside the set, then set stability must also be less than one. The staticness of interior (i.e., non-perimeter) states, however, cannot directly effect that set's stability measure.

6.1.1 Turbulence

On the other end of the spectrum from stability and staticness are measures of how likely the system is to change states. Because all these measures are defined in terms of probabilities, simple measures of dynamism can be calculated as one minus the appropriate measure above. There is one additional measure to present here; it is a measure of how predictable system dynamics are.

Definition 46 The *unweighted turbulence* of a **set** is the percentage of mathematically possible transitions that are actually observed (represented in the Markov model). A fully connected set of $|\mathbf{S}|$ nodes has $|\mathbf{S}|^2$ edges, so using k_i as the out-degree of S_i, turbulence is calculated as

$$\frac{1}{|\mathbf{S}|^2}\sum_{S_i \in \mathbf{S}} k_i.$$

This is equivalent to an individual-based measure that can be useful when you only know the average degree of the nodes in the set: the average of the percentage of states that the set's states can transition into. We can calculate **S**'s turbulence with the average ratio of each state's degree to the number of states in **S**:

$$\frac{1}{|\mathbf{S}|}\sum_{S_i \in \mathbf{S}} \frac{k_i}{|\mathbf{S}|} \approx \frac{1}{|\mathbf{S}|} k_{\text{average}}$$

Because every node's exit probabilities must sum to one, every node must have $k_i \geq 1$, and therefore this measure ranges from $\frac{1}{|\mathbf{S}|}$ to one. The lower bound approaches zero as $|\mathbf{S}| \to \infty$, but it can normalized onto [0 1] by subtracting $\frac{1}{|\mathbf{S}|}$ from the numerator and denominator. A turbulence of one means that the set is fully connected (including self-transitions for each state). The idea is that when each state has only a few possible transitions then there are far fewer possible paths through the system dynamics. If each state can transition into many others, then (like with the common usage of 'turbulent') there is a great deal of uncertainty regarding how the system's dynamics will unfold. We can see the intuitive relationship between turbulence and predictability in Fig. 8.

Plotting the degree distribution of the nodes in a set would reveal the set's turbulence profile (which might be the whole system or just a portion). If one were to find something like a power-law distribution (where a few states have many transitions and most have just a few transitions), then the high-degree states would seem to

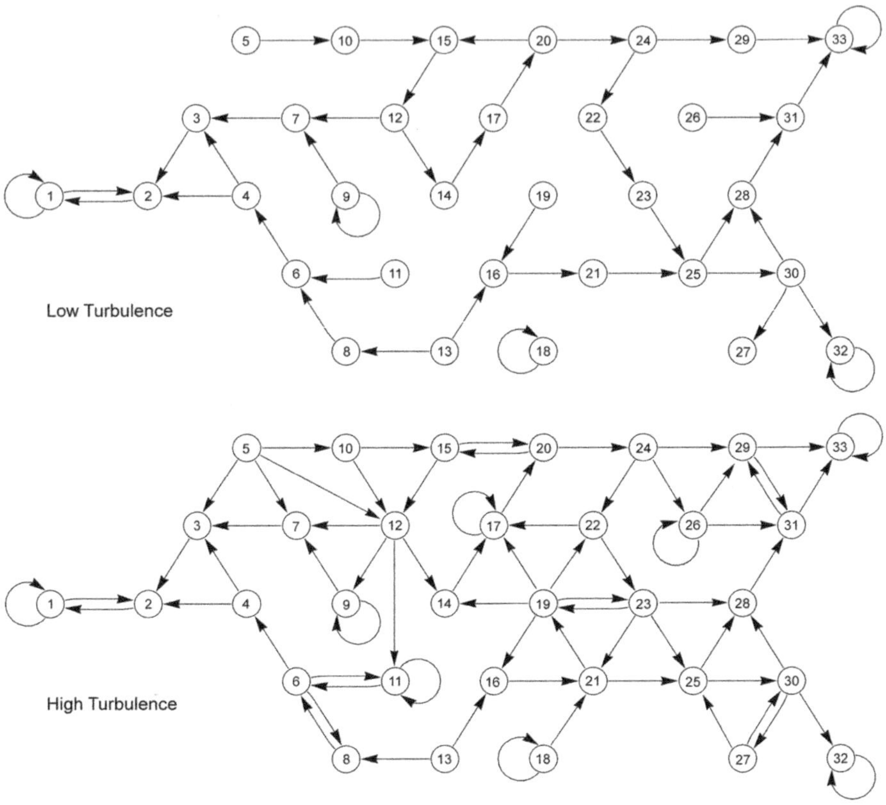

Fig. 8 The top Markov model has an unweighted turbulence value of 0.0386, and the lower one has a turbulence value of 0.0652. This provides an intuitive visual for how much easier the low turbulence system is to predict compared to the high turbulence one with the same collection of states and the same attractors

satisfy another concept related to tipping behavior. Combining turbulence profiling with (for example) the identification of perimeter states could be used to classify systems by their dynamical properties (see the Extensions section for more details).

Though the turbulence measure may provide sufficient information in many systems, it fails to differentiate the effects of high and low probability transitions. Transition weights clearly play a role in determining how likely it is that a particular trajectory will be taken rather than another. For example, if all the states in some set are such that all but one of each state's transitions have very small probabilities associated with them, then the set should be considerably less turbulent than if all the transitions are equally probable. We must use the edge weights to determine the effective degree of the nodes, but we cannot do so by simply summing the edge weights (as is typical in network theory) because the maximal effective degree occurs when all the edge weights are equal. One measure that achieves this feature is Shannon entropy, and another is the Simpson index; the latter one we adopt here.

Definition 47 The *effective degree* of state S_i with respect to set $\mathbf{S}$, denoted κ_i, equals

$$\frac{1}{\sum_{j \in \mathbf{k_i}} p_{ij}^2}$$

Although the effective degree concept/measure may be usefully applied to other networks, here we focus on its use for probabilistically weighted edge density.

To determine the weighted turbulence for some set $\mathbf{S}$ we need to combine the effective degree of each state with the systemic notion of what percentage of possible dynamics are actual. Because of the connection between degree used in Definition 46 and effective degree, we can simply replace the effective degree in the turbulence measure to calculate the weighted turbulence.

Definition 48 The *weighted turbulence* or just *turbulence* of the set $\mathbf{S}$ equals

$$\frac{\frac{1}{|\mathbf{S}|^2} \sum_{S_i \in \mathbf{S}} \kappa_i - \frac{1}{|\mathbf{S}|}}{1 - \frac{1}{|\mathbf{S}|}}.$$

This measure of dynamical uncertainty is normalized to between zero and one so that it can be usefully applied across domains. Naturally the location of the bin boundaries will affect the edge weights, and thus the measured uncertainty of the system. We can use this feature to explore binning parameters towards minimizing this uncertainty.

6.1.2 Uniformity and Entropy

We have two other related measures of dynamical uncertainty: uniformity and entropy.

Definition 49 The *uniformity* of the set $\mathbf{S}$ equals

$$\sum_{j=1}^{k_i \in \mathbf{S}} 1 - |P(\overrightarrow{S_i\ S_j}) - \frac{1}{k_i}|.$$

Because the sum of exit probabilities must sum to one, the average of the exit probabilities for state S_i is $1/k_i$ regardless of the value of k_i and the individual edge weights. The innermost component of this calculation finds the magnitude of the difference between each transition's weight and the average transition weight. We take one minus each transition's deviation from the average weight to calculate each transition's contribution to S_i's total effective degree. If all of state S_i's transitions have the same probability then the effective degree is k_i, otherwise it scales down according to the dispersion of its transition probabilities.

Both turbulence and weighted turbulence have an interesting and surprising double meaning. Because they measure how predictable a system's transition pattern is, and hence how much information about the system one would need to in order to predict the next transition and any future state, they act as alternative measures of *system entropy* (which can also be thought of as system uncertainty in some contexts). Tough the above formulations of system uncertainty (including Shannon entropy) hold promise for informing modeling decisions, more research is needed to evaluate them. Forthcoming work explores the differences among these measures of dynamical uncertainty.

6.2 Sustainable and Susceptible

In common parlance something is sustainable if it can perpetually maintain its operation, function, or existence. It is often used in connection to environmental considerations such as whether humans are using up resources faster than they can be replenished or to the ecological question of whether population dynamics will drive any species to extinction. Political institutions, academic reading groups, pools of workers, and any other system that undergoes inflows and outflows of its parts is a potential subject for sustainability considerations.

For this methodology, the dynamics of a system are captured as edges among the set of observed states, and so for this analysis a subset of states is sustainable if the system can *stay* within that subset of states. Because one identifies sets based on aspects of the states or a system-level property, this notion of sustainability can be applied to measure the maintenance of properties in diverse ways. Variations on the calculations produce varying measures of staying within a set, and each reports a slightly different concept of sustainability.

As a crude approximation to the long-term sustainability presented later, we can find the cumulative sum of the tth power of the set stability measure from Definition 44 up to some sufficiently large T (see Observation 6.1 below). We could call such a measure *naive sustainability* and identify conditions for its appropriate application, but instead we will move on to a more sophisticated measure. The previous measure is crude because within a set there may be (for example) heavily weighted cycles such that if the system starts in one of the cycle-states it is very likely to go around the cycle for a long period. To properly account for this, while still remaining agnostic over which state of $\mathbf{S}$ the system starts in, we calculate a refined sustainability measure.

These measures capture the flow of probability mass through the system and/or set of states. For the definitions we can consider the results at the limit of an infinite number of iterations, but in practice we need a way to know when to stop the calculations. One way is to introduce a parameter for the analysis run time T for the number of iterations the Markov process is run through to calculate the measures. Determining an appropriate value for T requires many considerations, most of which are outside the scope of this treatment, but "sufficiently large" is the standard phrase. The parameter T can often be replaced with a *min-flow* level can stops the analysis when the change in any states probability mass is small enough. Notes are made below for those considerations for T and min-flow that are crucial to understanding the measurements.

Definition 50 The *sustainability* of **S** is the average cumulative long-term probability density of states that remain in the set, having started from each state within the set:

$$\frac{1}{|\mathbf{S}|}\sum_{S_i \in \mathbf{S}}\sum_{t=0}^{T} P(s_{t+1} \in \mathbf{S}|s_t \in \mathbf{S}).$$

The algorithm for calculating sustainability starts with a vector of ones for the states in the set and zeros for states out of the set. It then iteratively applies *this vector* to the whole adjacently matrix taken to the tth power up to $t = T$ (see Observation 6.1). This is done each time to clear out the probability mass outside the set so that it can't return. The result is the sum of the resulting vectors for the states in **S** each step. Understanding this calculation is likely to make understanding Definition 51 of susceptibility much clearer.

Observation 6.1 *If the chosen set does not contain attractor states (states that are an equilibrium or part of an orbit), then some probability density "escapes" the set each iteration, and as $T \to \infty$ the internal probability goes to zero. For the algorithmic approach taken here, we utilize the fact that there exists some time T after which the remaining probability density in each state of* **S** *is less than any arbitrarily chosen minimum resolution (min-flow). If* **S** *contains an attractor then the probability density will never drop below any reasonable minimum resolution, but for some sufficiently large T the dynamics will have settled to nearly zero outside and some constant level inside the attractors. The measure as written is therefore only appropriately applied to sets not containing attractors, but accommodations can be made for attractors as well. In practice, T can be adaptively chosen as the number of iterations at which the change in the measure across Δt iterations (another parameter) is below some chosen accuracy parameter (which is a function of T). The exact value of T is unimportant and different values can be used for different performances of the analysis as long as the value of T is sufficiently large that the dynamics have settled.*

The resulting cumulative probability density values are difficult to interpret by themselves, but the normalized unit is a value that can be compared for any property/set for any system for any application of the analysis technique: it is domain-, scale-, and resolution-agnostic. Though due to the summation across time it is not a probability measure, it is a kind of likelihood measure of our expectation that the system will be in **S** given it starts somewhere in **S**.

Example 6.1 If **S** is chosen to be an energy plateau then all exit transitions are one-way. If there are no cycles in **S** then the probability mass will quickly dissipate from within **S** yielding a very low sustainability measure. If there is at least one core in **S** then there is a chance that the system will stay within the set up to any arbitrary time, but since that cycle cannot be an attractor the system's dynamics will leave the chosen set given enough time. The stronger the weights of the transitions among the cycle states the greater the sustainability measure.

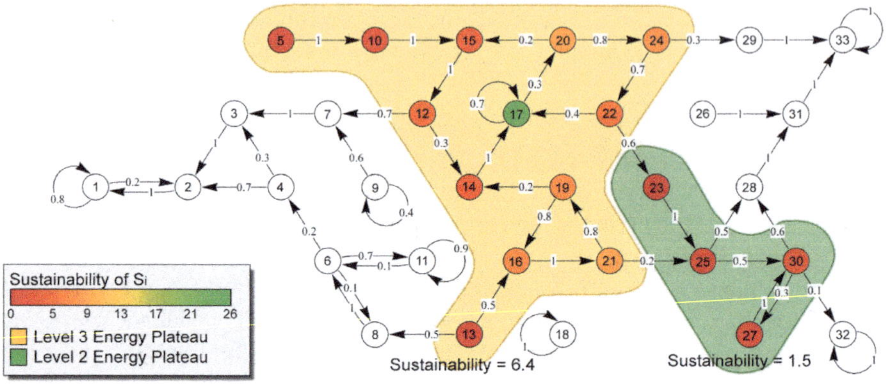

Fig. 9 Sustainability measures for two energy plateaus and their states of the running example system. Nodes are shaded by their individual sustainability values, and set sustainability is the average of these values across the nodes in the set. The energy-level 3 plateau contains several cycles, some of which have heavy weights, and it has a sustainability value of 6.4. The energy-2 plateau has just one cycle and most states cannot be revisited, and as a result its sustainability is 1.5

The energy plateau application is especially helpful for the "keep our options open" situations and where each equilibria is a different form of system failure (see Fig. 9). It can also be useful to calculate the sustainability of a basin of attraction excluding its attractor—which is also an energy plateau. A system may exhibit interesting and long-lived behavior within a basin of attraction that may reveal much more about the processes affecting a system than just which attractor it is likely to end up in. The time-to-equilibrium may be on the galactic time-scale and knowing what happens in the meantime will alter our understanding of the system's characteristics.

The term 'susceptible' is typically followed by 'to' and an indication of what the thing is susceptible to. I preserve that usage with the measure presented here. We will talk of sets as being susceptible, and states are what sets are susceptible to. This may sound odd, but the probability of transitioning out of a set depends on which state within the set the system is currently in. And the probability of losing whatever characteristic the chosen set of states represents is the probability of exiting the set from the state in question. In a particular usage, we would probably identify the aspect of the state that changes out of the set (the lever) as the thing that the characteristic is susceptible to, but that doesn't change the definition.

Definition 51 The degree to which $\mathbf{S}$ is susceptible to S_i is how much more (or less) likely it is to transition out of $\mathbf{S}$ conditional on it being in a particular state S_i of $\mathbf{S}$ compared to the sustainability of $\mathbf{S}$ overall.

$$\sum_{t=1}^{T} P(s_{t+1} \in \mathbf{S} | s_t \in \mathbf{S} \text{ and } s_0 = S_i) - \text{sustainability of } \mathbf{S}$$

Given this definition we can see that a positive susceptibility means a lower probability to stay within **S**.

Rather than single states, we can also determine and measure the set's susceptibility to the lever points. Recall that the *lever points* of an aspect are all those state transitions that result from a change in that aspect. We can calculate the susceptibility of **S** to a collection of lever points for aspect X_h by replacing the following formulation for the conditional part:

$$\sum_{t=1}^{T} P(s_{t+1} \in \mathbf{S} | s_t \in \mathbf{S} \text{ and } \overrightarrow{s_0\, s_1} \in \bigcup_{\overrightarrow{S_i\, S_j}}^{\mathbf{E}} X_{h(i)} \neq X_{h(j)}).$$

Example 6.2 The sustainable/susceptible analysis can be used to help systems maintain a performance level. We can take the set to be a contiguous collection of states that count as functional in some system: such as all the configurations of an airplane that the autopilot can manage. For the airplane system, some state changes will be exogenous perturbations due to environmental factors (wind, rain, pressure, lightening, passenger movement, etc.), others will be endogenous control adjustments by either the pilot or the autopilot, and some will be a mix. First, one would calculate the sustainability of the whole set of autopilot capable states. Then one would calculate how susceptible that set is to each state (or smaller collection of states). Using this information the autopilot and/or pilot could select actions that minimize susceptibility across the states visited and this means maximizing the probability of staying within the set of autopilot capable states. This example can be generalized to any case where maintaining functionality is the modeler's goal.

6.3 *Resilient, Fragile, and Recoverable*

Stability, Staticness, and Sustainability are different ways to measure a system's dynamics' tendency not to leave a state or set; we now turn to measures of returning to a state or set once it has been left.

Definition 52 A **state's** *resilience* is the cumulative probability of returning to that state given that the system starts in that state. The resilience of S_i equals

$$\sum_{\widetilde{S_i\, S_i}} P(\widetilde{S_i\, S_i}).$$

If the calculation is restricted to distinct paths, then this measurement is equivalent to

$$\sum_{t=1}^{T} P(s_t = S_i | s_0 = S_i).$$

This measure calculates the sum of the individual probabilities of returning in $1, 2, 3, \ldots, T$ time steps. Because the sum of exit probabilities of every state equals one and the probability of traversing a path is the product of the states along the path, this cumulative sum is always less than or equal to one and is a true probability measure.

Because were are not trying to measure the long-term probability of being in S_i here, the probability flow is eliminated once it returns to S_i. So in this case T is the length of the longest path from S_i back to S_i, which can be at most N. If the paths are not distinct, then states in a cycle will continue to leave and return infinitely. The restriction to distinct paths has the effect of stopping the measurement once the state has been returned to. This way it measures just the probability of returning to, rather than the probability of being in, that state.

Definition 53 A state's *fragility* is a measure of how likely it is that the system will never return to that state. This is just one minus the resilience of that state.

So equilibria have zero fragility and states with no return paths have a fragility of one. Measuring the degree of fragility requires the same calculation as measuring the resilience, but finding out whether a state is ever revisited is much easier because we can utilize Definition 18 of a state's reach (or out-component).

Definition 54 A state S_i is *brittle* if and only if it has a fragility value of one (i.e., a resilience value of zero). Brittle states are the ones such that

$$\forall h \; P(s_{t+h} = S_i | s_t = S_i) = 0.$$

Except for the brittle states which have a specific formal significance, the choice of whether to use a resilience or fragility measure will depend on which feature the user would like to highlight (glass half-full or half-empty). The brittle states of our example system can be seen in Fig. 10.

We can also define the resilience and fragility of a **set** in an analogous way.

Definition 55 *Set resilience* is the probability that the dynamics will return to a set if the initial state of a sequence is within the set.

$$\sum_{t=1}^{T} P(s_t \in \mathbf{S} | s_0 \in \mathbf{S})$$

Though the definition is exactly parallel to the single-state case, the algorithm to calculate this probability is considerably more difficult.

Observation 6.2 *A few facts about entering and leaving sets will help refine our understanding.*

(a) Transitions exit $\mathbf{S}$ *at the perimeter states* $\mathbf{S}_{out}$ *of* $\mathbf{S}$.
(b) Transitions enter $\mathbf{S}$ *at a set of entry points* $\mathbf{S}_{in}$ *of* $\mathbf{S}$.
(c) We can refine the definition of set resilience to $\sum_{t=1}^{T} P(s_t \in \mathbf{S}_{in} | s_0 \in \mathbf{S}_{out})$.

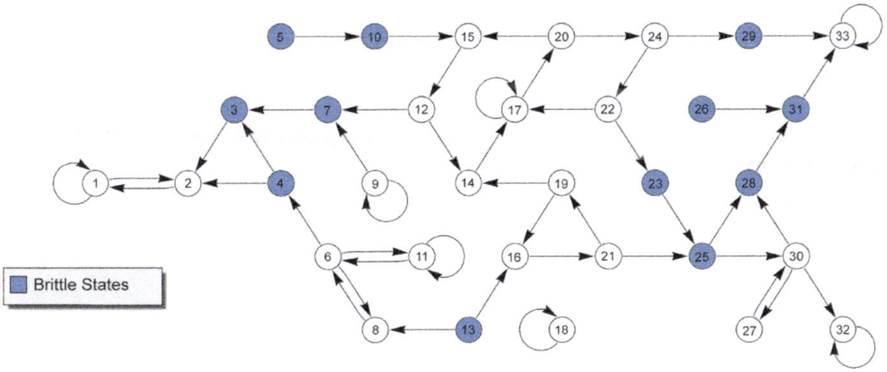

Fig. 10 A system's brittle states have no re-entry. Certain sets of brittle states can be considered for collapse or elimination from the Markov model (S_5, S_{10}, and S_{26}), whereas others are important placeholders

So to calculate set resilience we need first to find all the paths from each element in $\mathbf{S}_{out}$ to each element in $\mathbf{S}_{in}$. In the worst case this can be done in $O(|\mathbf{E}| + |\mathbf{N}|)$ time via a breadth-first search. Set fragility is one minus set resilience.

Theorem 6.2 *If* **S** *is an energy plateau then the resilience of* **S** *is zero.*

Proof. By Definition 38 of an energy plateau, it contains all the states in the system with the same number of reference states in their reach. Any transition out of such a set would be to a state with a different energy level, and as shown by Theorem 5.2 it must be a lower energy level. Also by that theorem, **no** transition can be to a higher energy level. Hence if a system transitions out of an energy plateau then it can never transition back into it. If the system cannot transition back into the set **S** then by Definition 55 **S**'s resilience is zero.

Above we saw how the susceptibility measure captures how much sustainability changes depending on the specific starting state. Next is a similar measure for resilience that reports how likely the system dynamics are to return to a set given that it exits via a particular transition.

Definition 56 A transition out of the set is *recoverable* to the degree that the system will return to the set after the transition. **S** is recoverable from $\overrightarrow{S_i\, S_j}$ to the degree calculated by

$$\sum_{t=2}^{T} P(s_t \in \mathbf{S} | s_0 = S_i \in \mathbf{S} \text{ and } s_1 = S_j \notin \mathbf{S})$$

Note that leaving via a particular transition is the same as exiting due to a particular lever change. Thus we can uncover the recoverability of a set of lever points for a particular aspect as the average of the recoverability of each transition in it. Also note that there may be multiple paths from S_i back into each $\mathbf{S}_{in}$ of $\mathbf{S}$. Each path leading from $\mathbf{S}$ back into $\mathbf{S}$ can be called a *recovery path*.

Example 6.3 Continuing with the autopilot example, imagine that there are many known points of failure for maintaining autopilot control. Each of these is a transition out of the set $\mathbf{S}$ via a known lever change. But not all failures are equally as problematic. By calculating the recoverability of each of the failure transitions they can be ranked by their seriousness. Such a ranking can guide both the pilot in adjusting to the failure and the autopilot to prioritize avoiding those transitions in the first place. Again, the autopilot example can be generalized to the maintenance of any system: political regimes, sports clubs, ecosystems, viable crop production, etc.

6.4 *Reliable, Robust, and Vulnerable*

Sustainability measures the likelihood of staying in a set given that it starts within that set, and resilience measures how likely it is to return to the set if the dynamics leave the set. The next step is to combine the features of sustainability and resilience to measure the probability of *being* in a set regardless of the trajectory. None of those measures includes the case where the system's state starts outside the set and then enters it. Even when the set of interest is an energy plateau (and resilience is zero by Theorem 6.2) the set may receive probability mass from parts of the system with higher energy levels. And in cases where a non-equilibrium analysis is appropriate, we might be comparing different subsets within an energy plateau. We now address these dynamics with more formal measures.

Definition 57 The *reliability* of a set is the average cumulative long-term probability density over the states in the set given that the system starts within that set:

$$\frac{1}{|\mathbf{S}|}\sum_{S_i\in\mathbf{S}}\sum_{t=1}^{T} P(s_t \in \mathbf{S}|s_0 \in \mathbf{S}).$$

This measure combines the concepts of sustainability and resilience, but it is not just the sum of those two measures. Reliability starts the flow in the set and calculates the probability of being in each state on each consecutive time step. It restricts the probability density summation to the specified set, but it tracks probability flow throughout the entire system. The reason that this isn't merely a sum of resilience and sustainability is because when combining those two it is not possible to track probability mass that leaves the set, cycles back into the set, and then circulates within the set (and maybe even repeats this process). With reliability we can reincorporate probability flow that leaves and then re-enters the set. A characteristic of the system captured by the set chosen is therefore reliable if it can be maintained or, if lost, can be regained.

Because reliability includes sustainability-like dynamics, it also requires special consideration if S includes attractor states. The difference again is that when S doesn't contain attractor states the probability mass of the states in S at time t approaches zero as $T \rightarrow \infty$. If S does contain attractor states then this mass approaches cT where c is a constant representing the probability mass store of the attractor states. For the current treatment one must omit attractor states when calculating reliability, but the full methodology includes techniques to also measure and compare sets that include attractors.

Theorem 6.3 *If* **S** *is an energy plateau then* **S***'s reliability equals* **S***'s sustainability.*

This theorem does not follow directly from Theorem 6.2 because there is no direct link between resilience and reliability, but the reasoning is the same. Because there cannot be any paths leading out of an energy plateau back into it, all the probability mass that **S** gets for the reliability measure is from the initial distribution. Leaving mass never returns, so that produces an equivalent measure as not counting the returning mass: this is the sustainability measure. Hence the sustainability values in Fig. 9 are also those energy plateaus' reliability values.

6.4.1 Robustness

Finally we add to the reliability measure the possibility that the system did not start in the set, but transitions into it.

Definition 58 The *robustness* of a set is the average cumulative long-term probability density over the states in the set, given that the system may start at any state.

$$\frac{1}{|\mathbf{S}|}\sum_{S_i \in \mathbf{S}}\sum_{t=0}^{T} P(s_t \in \mathbf{S}|s_0 \in \mathbf{N})$$

Robust characteristics not only have high retaining power and recoverability, they also draw the system in from states outside the characteristic set. Sets with high robustness values are sets that the system's dynamics tend towards. That description makes robust sets sound a lot like attractors; and this is as we would expect. Attractors will typically have high robustness measures on account of their perfect sustainability and the fact that typically several states will lead into them. The attractor-like behavior related to robust sets provides interesting and useful insights into many systems' dynamics.

Example 6.4 Sets that behave like (and are sometimes defined as) equilibria in other modeling techniques may be revealed to be highly robust non-attractor sets under the current analysis. The phenomena of *punctuated equilibria* describes a system that spends long periods of time in characteristic patterns with interspersed and short-lived periods of rapid change. In the Markov model representation we might see a subset of an energy plateau with multiple highly robust cycling subsubsets (a mantle

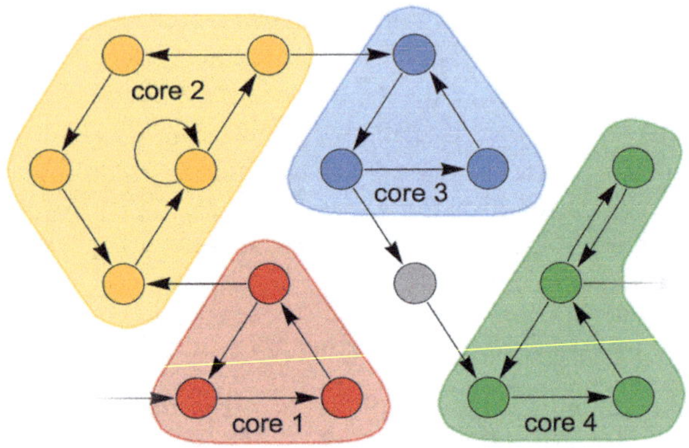

Fig. 11 Punctuated equilibria

with multiple cores by Definition 19, see Fig. 11). These cores could have relatively short transition paths between them and greater probabilities of cycling within than transitioning out of each core. Each core could appear as a distinct cohesive behavior pattern with a long-maintained property or characteristic. But because these cores are not attractors the system will eventually transition out of them and into the next core (or other dynamics).

Example 6.5 One of the foci of complex systems science is the study of the self-maintaining (or *autopoietic*) nature of dissipative structures. Dissipative structures are those where a continual flow of energy, matter, or other resource is necessary to maintain system structure and performance. Biological systems are like this, constantly changing and adapting to maintain functionality, and so are many other complex systems. These are systems where there are no point attractors[9] or all equilibria are states to be avoided so that the energy level of the system remains mostly constant. Some set(s) of states are preferred to others for exogenous reasons (functionality, performance, diversity, longevity, or other utility measures) and the goal is to maximize time spent in the desired states. The goal might also be to maintain some characteristic feature of transient system behavior. The current techniques offer new measures of behavior for non-equilibrium analysis. These can be used to embed an existing equilibria model into a larger context and/or to push down the level of analysis to see what is happening inside an "equilibrium".

6.4.2 Vulnerability

As before, we want to measure variations in the sensitivity of the property associated with the chosen set contingent upon other features of the dynamics. Using a definition parallel to that of susceptibility, the following measure calculates how much more (or

[9]There must be at least one attractor per system, but that attractor may be an orbit consisting of every state in the system.

less) likely the system is to be in a specific set, conditional on the dynamics starting in some particular state (not necessarily in that set).

Definition 59 A set's *vulnerability* is the difference in the average long-term probability density over the states in the set compared to the density generated by starting in S_i

$$\left(\sum_{t=1}^{T} P(s_t \in \mathbf{S} | s_0 = S_i)\right) - \text{robustness of } \mathbf{S}$$

This definition can be expanded to include the vulnerability of a set to another set of states in the obvious way.

The measures presented in this treatment are only a portion of the full breadth of the technique. Several other variations of the included measures, as well as more theorems and corollaries of these variations, have been developed. Measures from network theory such as betweenness, clustering, and community structure also have interpretations for the Markov model representation of data. For a more detailed treatment of the foundations of the methodology see [2] and related work (including software) currently under development. Some specific areas of focus are covered in the final section.

7 Extensions

The methodology presented in this chapter can capture a wide variety of system dynamics and provide the insight necessary to understand features key to analyzing complex systems, but certainly there is more work to do in refining, distinguishing, and improving these definitions. There are other dynamical properties in need of definition, and a suite of improved algorithms to run the analyses on actual data must be developed. Good methodology exists as a facilitator to good science, so the first and perhaps most important extension of this project is to apply these measures to models and datasets connected to substantive research projects. Doing so is essential to determining when this analysis is most (and least) insightful and to get feedback to improve and expand the methodology. All of these are in progress, and further collaboration is welcome. Descriptions of some specific extensions to this methodology immediately follow.

7.1 Non Probabilistic Definitions

There are two potential non-trivial objections to the above-given probabilistic accounts of properties of system dynamics. The first is that probabilistic definitions are inadequate because we aim to understand these features as properties that systems possess rather than dynamics they *might* have. As long as the above definitions reveal useful distinctions and patterns of system behavior the project is a success, but different and also useful measures may be available if built from a different formal foundation. I will, naturally, continue to pursue other and hugely different measures of system dynamics and strongly encourage input from others working on similar projects.

The other objection to the probabilistic definitions provided is that a person may insist that for many of these concepts the definition is incomplete without the causal explanation for how it comes about. Like all other statistics-like approaches these measures may be realized by many different micro-level dynamics. Some of those dynamics may not seem proper candidates for (say) robustness even if the data they generate reveals it as such from this analysis.

If somebody were to capture behavior that measures as highly robust with the measurement here, but fails to be robust for conceptual reasons, then I would consider the project an overwhelming success. And similarly for behavior that clearly fits the robustness concept, but requires a different mathematical treatment. These realizations would be akin to discovering scale-free degree distributions in many different networks from disparate research fields. In the case of power law degree distributions, finding that common network property urged researchers to pursue more deeply the phenomena, and they eventually uncovered several different mechanisms by which a scale-free network may be created. Our understanding of each of those systems greatly increased because we had a common yardstick with which to measure them. The probabilistic measures presented here are not intended to replace or make unnecessary the deeper scientific analysis—they are supposed to foster it.

7.2 Equivalence Classes for System Dynamics

An oft-cited goal of complex systems research is to identify common underlying mathematical properties in a myriad of seemingly very different phenomena. The Markov modeling technique allows us to create a common representation of almost any system's dynamics. Differences in the definitions of system states, however, will still mask many of the system similarities. That difficulty notwithstanding we can make great gains by identifying network *motifs* (repeated patterns in the graph structure) and establishing cross-disciplinary equivalence classes of system behavior. Achieving this goal will require solving issues with the choice of system resolution and "playing with" the resolution to find the matching patterns. Though this may sound suspicious, changing the resolution is nothing more than altering the level of organization to which we are applying the properties. As long as we are consistent in our application of these techniques then we may be able to discover similarities in many complex systems' dynamics. Any technique as complicated, sophisticated, and general as the one presented here is likely to require experience and expertise to apply appropriately.

The potential gains from identifying equivalence classes in system dynamics is huge. This would allow researchers to bootstrap understanding gained from a previously studied system by discovering that another system exhibited very similar dynamics. By establishing categories of system dynamics and sorting models/processes by their dynamical properties, our ability to make progress going forward is quickened and simplified. Complex systems is supposed to be a field focused on processes, but our ability to manage data regarding dynamics is still in its infancy. The measures presented here are supposed to help, using them to form equivalence classes that usefully categorize dynamics would help significantly more.

7.3 Metastatistics

I have said elsewhere in this paper that I consider the methodology presented here to be similar in kind to statistics. It starts with data (perhaps generated from simulations), fits a model to the data (a Markov model), and then purports to *describe* the original system with measures over that model (my definitions). Statistics as we usually see it utilizes a different type of model (some form of distribution or estimator) to represent a phenomena or system, but its purpose and general method of analysis are very similar. My Markov modeling procedure is clearly different from other sorts of models in that it is not used to explain the phenomena being analyzed, but merely describe it—a feature it shares with statistics. Both standard statistics and my methodology can produce evidence that some generative theory-driven model does explain the observed data, but the theory behind the generative model is what is doing the explaining.

Standard statistics and my methodology are certainly not unique in their abilities to describe and measure, but not explain, phenomena. Much of complex network analysis can be seen in this light as well. The network representation facilitates the calculation of measures on the generating data but not because the links identified in the network representation are in the actual system's features. Classifier systems, Bayes nets, hidden Markov models, and neural nets are all further examples where the formal representation can permit measures and produce predictions without mirroring the structure of the underlying behavior-generating system. Seeing all these different techniques under the same metastatistical light may allow us to (1) bridge gaps among these techniques, (2) identify broader guidelines for the proper application and interpretation of these techniques, and (3) find new statistics-like techniques with additional desired features.

8 Conclusions

The definitions of tipping- and robustness-family system dynamics presented here are meant to draw clear boundaries around distinct, yet often confounded, concepts relating to the changing or maintenance of some system property. Using a specially created Markov model representation of system dynamics called the *Revealed*

Dynamics Markov Model (or RDMM), researchers from disparate fields can capture model behavior in a unified mathematical object. Applying the above probabilistic definitions to this mathematical object produces a measure of each of the distinct tipping- and robustness-family concepts. Thus the dynamical properties of very different systems can be analyzed and compared in a rigorous fashion. This methodology is meant to facilitate research in any domain for which understanding features of processes are important, and also to provide a baseline for more integrative multi- and inter-disciplinary research and analytics.

Although the intended primary application is to analyze simulation data, and especially data from agent-based models, the technique is not limited to computer-generated data. We have also applied this methodology to empirical data including primate behavior, patterns in renewable energy use/production, neural signals, virtual and world politics to name a few. These applications have pushed refinements in the binning methods, analysis methods, and the range of measures beyond what is covered in this foundational and very general work. Although still very much in its infancy, the approach described here continues to grow and reveal its strengths and weaknesses.

References

1. Bednar, J.: The Robust Federation: Principles of Design. Cambridge University Press (2009)
2. Bramson, A.L.: Measures of tipping points, robustness, and path dependence (2008). arXiv:0811.0633v1
3. Bramson, A.L.: Formal measures of dynamical properties: tipping points. AAAI Technical Report, vol. FS-09, pp. 2–11 (2009)
4. Clauset, A., Moore, C., Newman, M.E.J.: Hierarchical structure and the prediction of missing links in networks. Nature **453**, 98–101 (2008)
5. Farrell, B.: Political science . The Nation. (February 13, 2006). http://www.thenation.com/doc/20060227/farrell
6. Fontana, W.: The topology of the possible. Technical Report, Santa Fe Institute (2003)
7. Gladwell, M.: The Tipping Point. Little, Brown, and Company (2000)
8. Granovetter, M.: Threshold models of collective behavior. Am. J. Sociol. **83**, 1420 (1978)
9. Grodzins, M.: Metropolitan segregation. Sci. Am. **197**(4), 33–41 (1957)
10. Lempert, R.: A new decision sciences for complex systems. Proc. Natl. Acad. Sci. **99**(Suppl 3), 7309–7313 (2002)
11. Lint, J.V., Wilson, R.: A Course in Combinatorics. Cambridge University Press, Cambridge (1992)
12. Rubineau, B., Fernandez, R.: Tipping points: referral homophily and job segregation (2006)
13. Schelling, T.C.: Dynamic models of segregation. J. Math. Sociol. **1**(2), 143–186 (1971)
14. Walker, G.: The tipping point of the iceberg. Nature **441**(7065), 802–805 (2007)
15. Wolf, E.P.: The tipping-point in racially changing neighborhoods. J. Am. Inst. Plan. **29**(3), 217–222 (1963)
16. Young, H.P.: The diffusion of innovations in social networks. In: Blume, L.E., Durlauf, S.N. (eds.) The Economy as a Complex Evolving System, vol. III. Oxford University Press (2003)

Identifying Unexpected Behaviors of Agent-Based Models Through Spatial Plots and Heat Maps

Christopher J. Lynch, Hamdi Kavak, Ross Gore and Daniele Vernon-Bido

Abstract Verification and validation (V&V) techniques commonly require modelers to collect and statistically analyze large amounts of data which require specific methods for ordering, filtering, or converting data points. Modelers need simple, intuitive, and efficient techniques for gaining insight into unexpected behaviors to help in determining if these behaviors are errors or if they are artifacts resulting from the model's specifications. We present an approach to begin addressing this need by applying heat maps and spatial plots to visually observe unexpected behaviors within agent-based models. Our approach requires the modeler to specify hypotheses about expected model behavior. Agent level outputs of interest are then used to create graphical displays to visually test the hypotheses. Visual identification of unexpected behaviors can direct focus for additional V&V efforts and inform the selection process of follow-on V&V techniques. We apply our approach to a model of obesity.

Keywords Verification and validation · Heat map · Spatial plot · Visualization · Agent-based models

1 Introduction

The macro level behaviors that a given agent-based model (ABM) is intended to recreate may be well understood but the mechanisms that lead to these behaviors may not be known [3, 30]. Simultaneously, some expected behaviors and types of interactions among the agents within that system may also be known or assumed to be understood. However, the modeler may face (1) incorrect, incomplete, or contradic-

C. J. Lynch (✉) · H. Kavak · R. Gore · D. Vernon-Bido
Virginia Modeling, Analysis and Simulation Center, Old Dominion University, Suffolk, VA 23435, USA
e-mail: cjlynch@odu.edu

T. Carmichael et al. (eds.), *Complex Adaptive Systems*, Understanding Complex Systems, https://doi.org/10.1007/978-3-030-20309-2_6

tory knowledge about the system; (2) disagreement among the model's stakeholders on how to address the problem or over which problem to solve [19]; (3) challenges resulting from conflicting model characteristics [18]; (4) challenges resulting from simplifying the set of allowable agent behaviors into a finite set for implementation and execution; (5) challenges in examining the combinations and sequences of events that lead to various system level outcomes [11]; and (6) challenges due to behaviors of model components changing over time [25]. Therefore, an ABM serves as a theory tester to see if the agent and environment specifications produce any *unexpected behaviors*.

Unexpected behaviors are model outcomes that cannot be explained by the model's specifications and do not pass the selected V&V techniques, such as making comparisons against trusted simulations in the same application domain, data sets from physical experiments, or subject matter experts' (SMEs) opinions [30]. It is the modeler's duty to determine if any of these behaviors represent errors within the model. Ultimately, these unexpected behaviors require understanding and explanation to determine if the behavior is an error or new knowledge in the application domain [1]. However, as an initial step, the occurrences of these behaviors first need to be identifiable from simulations' executions.

Verifying and validating unexpected behaviors requires accumulation of insight and understanding into the behaviors and the conditions under which the behaviors arise. An unexpected behavior either becomes part of a set of behaviors one considers valid or it is deemed invalid. Visual V&V methods can assist in identifying unexpected behaviors by reframing the simulation execution into a variety of different views. Each view can be tailored for the intended audience such as the modelers, the model's stakeholders, and subject matter experts at varied levels of expertise based on their perceived level of knowledge about the modeled system. Effective visualizations can help to identify, communicate, and understand important behavior from the model while facilitating insight through the detection of key features within the output [15, 26, 27]. A survey conducted on simulation modelers found that the most common approach for verifying ABMs was through visual inspection with over one third of the respondents utilizing this approach [21].

Visualization can assist the modeler in providing clearer pictures on how and why the simulation is behaving in expected or unexpected ways and also assist in conveying these findings to interested parties. To this end, we present a technique for exploring the evolving conditions within an ABM through the use of spatial plots to examine agent positions and distributions and heat maps to examine agent density statistics. We provide a use case of an ABM to illustrate our technique.

2 Background

Visualization enhances the experience of model execution by graphically representing parameter levels, distributions of values, network relationships, and interdependencies [34]. This is accomplished in part by the ability to place the simulation into context by co-locating different types of data to facilitate exploration or gathering explanations within the simulation [29]. Visual techniques provide numerous opportunities for V&V of ABMs during execution. Animation, for instance, provides the modeler the ability to observe how the system functions and helps to communicate results [23]. Data can be displayed through information visualization to gather insight through a number of graphical techniques. Some of the more commonly used techniques include scatter plots, line plots, histograms, box plots [25], bar charts, icon-based displays, dense pixel displays, stacked displays [13], data maps, time-series plots [10, 32], polar diagrams [26] and cluster heat maps [35]. These techniques utilize datasets dependent upon size, type, or dimensionality for creating visual [13] representations.

Agent-based models are prime candidates to use spatial plots and heat maps as an analytic tool. Heat maps commonly assist in visualizing spreadsheets, data sets, and matrices [4] and are easily extendable for displaying the locations of agents or environmental objects as well as clearly conveying areas where agent groupings are occurring. Heat maps can represent interactions between agents with respect to geographic location (where x and y coordinates make up the x and y axes of the heat map) or use the agent types to form the x and y axes and utilizing the color scheme to represent interactions. The heat map representations allow for a simple and easy method for detecting if agents move through areas that they are not supposed to move through due to the model specifications. It also identifies interactions or communications between agents that are not allowed to interact. Similarly, it allows the modeler to observe if any expected communications or interactions do not occur.

Visualizations convey information about the simulation; as such, it is important that the selected visualizations do not distort the representation of model outcomes and decrease the usefulness of the model [5, 10]. When incorporating visual components for V&V, several challenges exist for visualizing model results, including representing model components, representing the magnitude of changes, and poor visualization choices adding an additional level of complexity to the model [33]. These challenges specifically apply to ABMs in the following ways:

- Challenge with misrepresenting model components: Showing agent locations without any additional information does not necessarily convey how or why the group formed. Visualizations that also convey internal agent attributes, such as age or movement behavior, can provide a more accurate view of what is happening to cause the grouping behaviors.
- Challenge with representing the magnitude of model changes: Visualizations based on altering agents' representative colors and local min/max provide the illusion that a min/max value always occurs during execution; however, this does not necessarily mean that an absolute min/max value occurred.
- Challenge with model complexity: Using the same colors, icons, or shapes for different elements within the visualization or using different time scales or metrics between visual elements make it difficult to interpret the results.

Technical challenges are also important for visualizing simulations, including usability, scalability, and integrated analysis of heterogeneous data [16]. Usability refers to designing visualizations that successfully contribute to the advancement and use of visualization research. The modeler is responsible for creating a visualization that model users can easily understand. Scalability refers to how well visualization tools apply to representing large data sets. In this case, too many overlapping representations within a single view area become indecipherable and fail to convey useful information. Integrated analysis of heterogeneous data deals with visualizing data obtained from various locations in various formats. All of the data being visualized needs to be clearly conveyed for each visualization and any differences in scale or magnitude should be explicitly stated.

We can potentially improve identification of unexpected behaviors by visualizing different perspectives of the same information. For instance, visualization of the spatial distribution of an agent population as an x-y plot, as a density function, or through their movement patterns can aid analysis. This provides an extra level of insight into the model's behavior and can help a wider variety of model users or stakeholder to get a better grasp of what the simulation does. However, the creation of multiple types of visualizations may require the collection of a greater amount of data that is periodically collected throughout the simulation runtime. This entails conducting *trace validation* which focuses on examining micro-level occurrences within a model to reveal errors [2]. Traces provide the behavior of agents or the model over time to determine if the logic is correct and the simulation produces believable values [9, 11, 14]. However, the large volumes of data created through trace validation is difficult to analyze and can be burdensome to interpret [6, 36].

There have been several efforts to provide SMEs with an interface to collect, analyze, and interpret traces. Xiang et al. [36] present a validation process for analyzing a natural organic matter model by using traces, graphs, charts, and model-to-model comparisons. Graphical charts test if the produced data curves match the expected distribution curves. This approach is suitable for inspecting the validation of population level statistics. Diallo et al. [8] provide a V&V Calculator for examining trace data and providing a statistical measure of model characteristics that are likely contributors to various model behaviors. Courdier et al. [6] use the Geamas Virtual Laboratory (GVL) tool to collect traces of a biomass ABM for analyzing animal wastes management. These traces collect (1) sets of messages exchanged between agents, or (2) a historical accounting of simulation execution per agent or group of agents. Visualization tools inspect these traces and identify interactions that lead to successful agent negotiations. Their visualization tool filters traces based on specific agents or characteristics. However, the GVL tool becomes unwieldy for analysis of traces once exceeding several dozen agents.

3 Methodology

Our methodology assumes that a simulation exists and is suspected to be correct. We propose a five-step approach for exploring the simulation for unexpected behaviors as shown in Fig. 1. Step 1 requires creating hypotheses, specifications, and

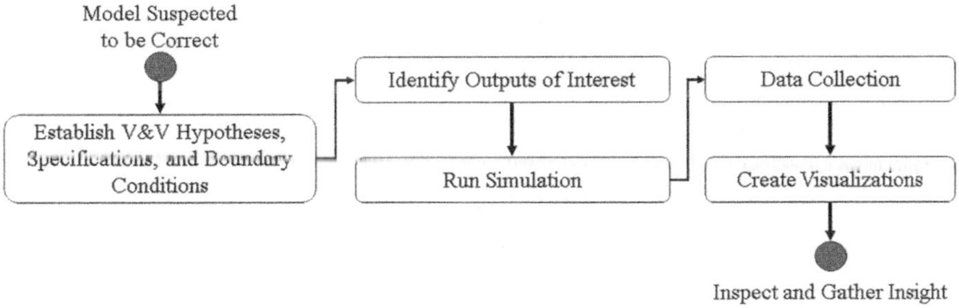

Fig. 1 Methodology for visually identifying unexpected behaviors of ABMs to gather insight

boundary conditions for establishing baselines for determining if outcomes appear unexpected. Step 2 involves identifying outputs of interest; i.e. the outputs required to check the hypotheses, specifications, and boundary conditions identified in Step 1. Step 3 involves running the simulation. Step 4 involves collecting the outputs of interest identified in Step 2 from the simulation runs. Step 5 involves creating visual representations of the outputs of interest in order to provide a means for exploring the outcomes. These visualizations are compared against the V&V hypotheses, specifications, and boundary conditions from Step 1 to identify any suspicious behaviors. We use visualization techniques in the form of spatial plots and heat maps to conduct V&V on the simulation.

Spatial plots represent agent locations based on their X-Y coordinates. Heat maps are two-dimensional data matrices that assign color values to each point in the matrix [35]. Higher values on the heat map display at different colors than the lower values in order to visually present the difference in values across the heat map's surface. We create heat maps using the exact dimensions as the agents' environment and simply count the quantity of interest at a particular location on the grid. A smoothing function is applied while transitioning from the high values matrix cells to the lower value adjacent cells. The smoothing function ensures color continuity; thus, improving the viewer's experience.

We utilize these visual methods to identify unexpected behaviors by reframing the simulation execution into a variety of different views, such as agent locations, movement patterns, or agent attribute values distributed across the environment to name a few. Each of these views can be tailored for the intended audience based on their prior knowledge of the system. To collect the micro-level data needed to identify errors we apply trace validation and store agent locations and parameter values over time. Then we create spatial plots to examine agent positions with respect to the environmental dimensions. We explore and examine population statistics and their distribution across the environment by creating heat maps.

4 Use Case: Obesity Simulation

We apply our methodology to inspect a simulation of obesity comprised of a society represented by a combination of individuals and environmental factors. We develop our model by first establishing a set of base assumptions and constraints to illuminate expected outcomes and boundary conditions for our obesity system following the Modeling and Simulation—System Development Framework [31].

4.1 Model Details

This is a simplified version of the Obesity Model described in [8, 17, 28] in which we focus exclusively on weight gain of individuals due to their access to food within their environment. The model consists of four agent populations: *people*, *homes*, *restaurants*, and *workplaces*. People have three possible eating habits: low, medium, or high calorie. They eat three meals a day at fast food, markets, and non-fast food restaurants [20, 22]. The number of calories gained from each meal is dependent upon the person's eating habit along with the restaurant type. The basal metabolic rate (BMR) equations capture the weekly calorie requirements for maintaining a person's current weight [12, 24] as shown in equations 1 and 2 (with Weight in pounds, Height in inches, Age in years, and BMR in calories).

$$BMR_{male} : 88.362 + (29.535 * Weight) + (1.889 * Height) - (5.677 * Age) \quad (1)$$

$$BMR_{female} : 447.593 + (20.386 * Weight) + (1.219 * Height) - (4.330 * Age) \quad (2)$$

The remainder of our core assumptions and constraints that form our model:

- *People* have age, height, weight, and gender attributes in order to use Body Mass Index (BMI) to classify as Obese (BMI $\geq$30), Overweight (30 $>$ BMI $\geq$ 25), Normal (25 $>$ BMI $\geq$ 18.5), and Underweight (BMI $<$ 18.5).
- *People* (1,000 created), *Homes* (100 created), *Workplaces* (100 created), and *Restaurants* (100 created) are uniformly distributed throughout the environment.
- The environment is a 625 $\times$ 625 unitless grid.
- Each person is assigned one *home* and one *workplace* at random.
- There is no limit on the number of people that can be assigned to a *home* or *workplace*.
- Starting age, weight, and height are specified by the user, but all *people* are at least 18 years old.
- *People* move from their home to their work, and back to their home (exclusively in this order) every day.
    - One meal is obtained at each location each day for a total of three meals.

  - To obtain a meal, each person selects a restaurant within 30 units of their current location at random.
  - Calories are determined based on the restaurant selected combined with the person's eating habit.
  - Calorie gain accumulates over the course of a week for each person.

- Restaurants produce a calorie amount for each meal based on the restaurant's type:
  - 400–900 calories per meal from Market type.
  - 600–1100 calories per meal from Non-Fast Food type.
  - 800–1500 calories per meal from Fast Food type.

- Eating habits modify the allowable calorie gain from each meal:
  - Low-Calorie eating habit pulls samples from the bottom 50% of the restaurant's calorie range.
  - Medium-Calorie eating habit samples within the 25–75% (middle) of the restaurant's calorie range.
  - High-Calorie eating habit samples from the top 50% of the restaurant's calorie range.

- People update their weights at the end of each week based on the difference (positive or negative) between their calories consumed that week and their current BMR value.
- Calorie gain resets to 0 each week and the obesity status of each individual is tracked over time.

We formulated two hypotheses to drive the selection of assumptions and constraints: (1) we hypothesize that the home and work locations of the people agents affect the growth of obesity over 15 years due to their proximity to different types of restaurants; and (2) we hypothesize that eating habits affect the growth of obesity over 15 years.

4.2 Experimentation

Creating spatial plots and heat maps of our outputs allows us to verify that the assumptions and constraints connected to the test cases are not violated and to conduct an initial validation of our two hypotheses. We utilize these visualizations to conduct a quick, visual search into the model's behaviors to identify obvious errors (such as an agent appearing outside of the expected 625 $\times$ 625 grid) while receiving intuitively analyzable feedback on what is happening within the simulation and what contributing factors appear to be supporting these happenings. We use the visualizations to gain insight into the model which can then assist the modeler in identifying next steps for additional verification and validation exercises. Note that while we display visualizations on the results of single runs to convey our approach, replications remain necessary for proper testing and each replication requires its own set of visualizations.

Hypothesis 1 suggests several items that we want to check within the simulation. On the verification side, we want to ensure that (1) all agent populations appear to be uniformly distributed within the 625×625 environment, (2) that no agents move outside the bounds of the environment, and (3) that people move between homes and workplaces every day. To assess the validity of our hypothesis, we identify the following outputs of interest:

- The number of calories that each person gains from each restaurant per meal;
- The location of each person's home and workplace;
- The location and type of each restaurant that the person visits per meal;
- Each person's weight/BMI classification over time.

We run the test case for 15 years of simulation time and collect the outcomes of interests for each agent. For this experiment, we equally distribute the three eating habits throughout the People population. We then create spatial plots to visually inspect the three verification items of interest by (V1) plotting each agents position within a spatial plot, (V2) plotting only agents which appear outside of the environment grid, and (V3) plotting People agents whose positions do not change throughout the day or whose end of day positions differs from their start of day positions. These three visualizations do not reveal any unexpected behaviors. The spatial plots for V1 appear uniformly distributed (at minimum they lack the appearance of large agent clusters) as expected and the plots for V2 and V3 are blank (indicating no violations occurring within the first test case).

We then create heat maps to spatially represent the weight density of people agents and density of high-calorie meals sold, as depicted in Fig. 2a, b. To create Fig. 2a, we count the number of People at each coordinate with each BMI classification using both home and workplace locations. Then, we compute the density of the obese population based on the number of people living in each home and working at each workplace. For Fig. 2b, we plot heat map of high-calorie meals sold within the environment to reveal the locations where people with high-calorie eating habits are eating. Figure 2c displays the heat maps' color scale with 1 representing high values and 0 representing low values. A comparison of Fig. 2a, b shows that there does not appear to be a visual correlation between the density of obese population locations and the number of high-calorie meals sold. This can be inspected by comparing the locations of the red areas between the two heat maps. The density of the high-calorie meals sold appears to be primarily centered within this simulation run, while the middle-right and middle-left positions within the environment contain a majority of the obese population. Additionally, the distribution of obesity appears to be largely stationed near the normal weight category (shown as white areas within Fig. 2a). This finding does not appear to directly support our hypothesis. However, there are several contributing factors that may influence this visualization, such as the fact that we created Fig. 2a using the home locations of the People agents, so a heat map of obesity distribution based on workplace location exclusively or a heat map based on an averaged obesity distribution based on home and workplace locations may

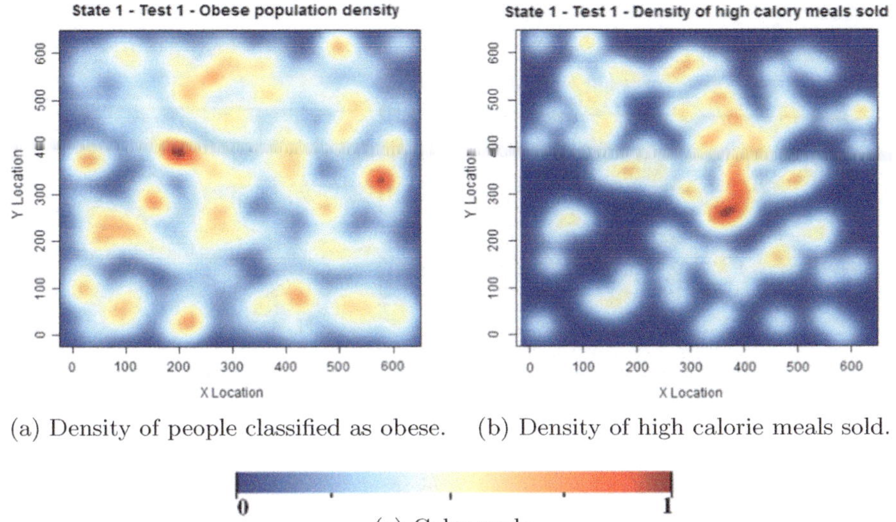

(a) Density of people classified as obese. (b) Density of high calorie meals sold.

(c) Color scale.

Fig. 2 Test case 1

provide better insight. After conducting and inspecting replications, our hypothesis does appear to hold.

For our second hypothesis, we expect eating habits to have a significant effect on the level of obesity after 15 years. For this experiment, we assign a low-calorie eating habit to all people. We maintain the same set of outputs of interest and verification checks from the previous case. In addition, we also want to confirm that none of the meals came from the top 50% of calorie distributions from each restaurant. Figure 3a displays a heat map of the density of obese people at the conclusion of the run while Fig. 3b displays a spatial plot of all of the restaurants' positions. An inspection of Fig. 3a quickly reveals that people are still able to become obese without eating high-calorie meals. Figure 3b shows that the distribution of restaurant locations appears to more closely resemble a uniform distribution than the first test case. A larger sample size is required to make a final determination, but this is a promising outcome. A heat map of high-calorie meals sold during revealed that zero was produced which successfully passes the extra verification check for this test.

From the verification perspective, people with healthy eating habits becoming obese does not contradict any specific assumption from the model design and does not necessarily constitute an error. However, this raises several questions with respect to the model's validation and reveals that this is an area which requires further exploration. This could indicate that the minimum and maximum values of the restaurants' calorie ranges are too high or low, that the calorie range adjustment for each eating habit (i.e. top or bottom 50%) is too large, or that the BMR equations are not updating correctly within the agents.

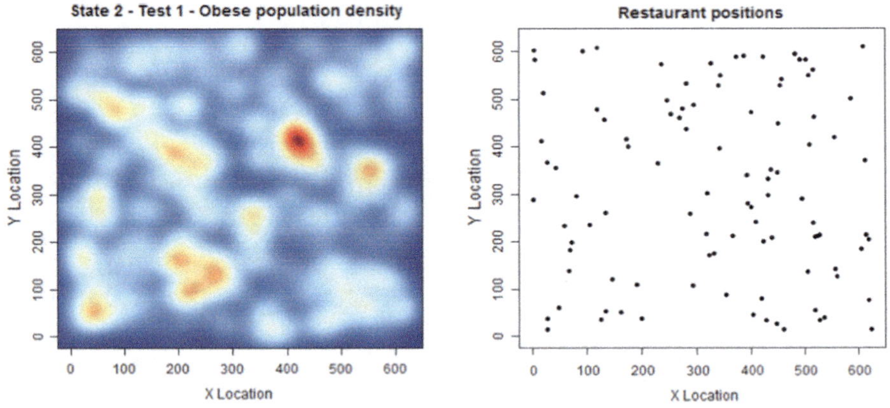

(a) BMI density of population. (b) Restaurant distribution.

Fig. 3 Test case 2

Ultimately, this issue results from an oversimplification of the model's assumptions in that we do not account for physical activity. By incorporating the BMR equations without modifications based on the activity, we have unintentionally constrained the model so that nobody is physically active. With this constraint taken into account, Fig. 3a is a believable outcome for the model. However, to bring the model in line with hypothesis 2, the model will have to be adjusted to account for physical activity and its effect on BMR and V&V will need to be re-conducted.

5 Discussion

Creating the mentioned plots only require obtaining each agent's x and y coordinates, the initial values of parameters pertaining to the outputs of interest (e.g., eating habit), and the final values of parameters pertaining to the outputs of interest (e.g., weight). Our methodology includes a number of limitations under which its use may not be feasible. As reported by [6, 36], visualizations become unwieldy for analysis when exploring dozens or more simulation runs at a time and that it is best suited for small sets of runs. While our use case focuses on the post-execution examination of single runs, heat maps can also be used at runtime to convey changes, illuminate clustering, or display variable distributions for the whole population. Additionally, these tests do not confirm that the model is completely error-free, they simply confirm that no errors were found with respect to the specific runs and output data involved with creating the spatial plots and heat maps. Additional tests are required for verifying assumptions or answering questions not directly associated with these tests.

A benefit provided by the use of heat maps is the ability to show the same data set under multiple perspectives, such as displaying the prominent areas of obesity along with the prominent locations of high-calorie meals sold. This allows for quick, informal comparisons to help determine if relationships appear to exist between agent types. Model constraints can be checked for violations or inconsistencies using visualization. In addition to heat maps and spatial plots, other visualizations, such as box plots and Q-Q plots, can further communicate differences between simulation results and the real system data for V&V [25].

As an initial V&V step in checking for unexpected behaviors, visualizations provide an intuitive interface for drawing inferences about the correctness of the simulation's execution and can help the modeler to select appropriate follow-on techniques. Our approach aids in the verification of some of the model's assumptions and constraints (those that pertain to the hypotheses being tested) while working to validate model hypotheses. Additional tests should still be conducted before formally accepting or rejecting hypotheses. The minimal background knowledge required to examine these graphics combined with the minimal data collection efforts needed from the simulation provides potential as a general approach for identifying unexpected behaviors within ABMs.

6 Conclusion and Future Work

We present an approach for identifying unexpected behaviors of ABMs that has the potential to enhance V&V practices for ABMs. We utilize a combination of spatial plots and heat maps to allow modelers to explore the interactions within their models. Heat maps allow for the creation of insight into the model based on visualization of the output data from the model. These explorations can assist the modeler in understanding how and why macro level behaviors are appearing within the simulation and guide the modeler in conducting additional tests. Spatial plots and heat maps are ideal for verifying and validating ABMs as they are suited for revealing the results of interactions between agents. We note that other types of visualization techniques can provide additional insights and can be further explored as future work in ABM V&V research.

Our use case conveys the ease of use involved with searching for unexpected behaviors of ABMs through visual inspection. Identifying assumptions, constraints, boundary conditions, and outputs of interest may be the most challenging aspect of our approach as it requires a high level of familiarity with the model. However, it does help to ease some of the knowledge requirements for V&V by removing the need to understand how to conduct statistical tests. The created visualizations assist in communicating how the model works, why certain outcomes occur, and increase confidence that the model's results are trustworthy. It remains the purview of the modeler to use proper visualization practices within these visualizations to convey the information in a manner that does not mislead its viewers. While we applied our approach to a snapshot of the simulation outcomes, these visualizations can also be applied during runtime to gain real-time insight into the dynamics of the agent behaviors.

Future work involves connecting our approach with existing V&V techniques to more efficiently explore the solution space and identify unexpected behaviors. This includes designing methods for automatically filtering the created spatial plots and heat maps to reduce the volume of graphs that need to be examined to only those containing potentially suspicious outcomes. For instance, [11] present an enhanced trace validation method for ABMs which facilitates SME insight without visualization by leveraging existing work in the field of statistical debugging. This method could be utilized in conjunction with visualization techniques to quickly isolate the causes of unexpected behaviors once identified. As a complementary approach to visualization, the use of sound can be explored as an option for revealing unexpected behaviors during runtime, such as ongoing work in [7]. Another avenue for future work includes an examination of what combinations of visualization techniques are best suited for identifying different classifications of unexpected behaviors, such as using heat maps to examine macro-level obesity behaviors while using Q-Q plots to compare distribution-based model assumptions against their actual distributions.

References

1. Axelrod, R.: Advancing the art of simulation in the social sciences. Complexity **3**(2), 16–22 (1997)
2. Balci, O.: Verification, Calibration, and Testing, book section 10, pp. 335–393. John Wiley and Sons, Inc., New York, NY (1998)
3. Bonabeau, E.: Agent-based modeling: methods and techniques for simulating human systems. Proc. Natl. Acad. Sci. U.S.A. **99**(Suppl 3), 7280–7287 (2002)
4. Chi, E.H.H.: A taxonomy of visualization techniques using the data state reference model. In: IEEE Symposium on Information Visualization, pp. 69–75. IEEE, Piscataway, NJ (2000)
5. Collins, A.J., Knowles Ball, D., Romberger, J.: A discussion on simulations' visualization usage. In: Yilmaz, L., Chan, W., Moon, I., Roeder, T., Macal, C., Rossetti, M. (eds.) Proceedings of the 2015 Winter Simulation Conference, pp. 2827–2835. IEEE, Piscataway, NJ (2015)
6. Courdier, R., Guerrin, F., Andriamasinoro, F.H., Paillat, J.M.: Agent-based simulation of complex systems: application to collective management of animal wastes. J. Artif. Soc. and Soc. Simul. **5**(3), 1–17 (2002)
7. Deuro, J., Lynch, C.J., Kavak, H., Padilla, J.J.: Incorporating sound in simulations. In: Chan, W., D'Ambrogio, A., Zacharewicz, G., Mustafee, N., Wainer, G. (eds.) Proceedings of the 2017 Winter Simulation Conference (2017)
8. Diallo, S.Y., Gore, R., Lynch, C.J., Padilla, J.J.: Formal methods, statistical debugging and exploratory analysis in support of system development: towards a verification and validation calculator tool. Int. J. Model. Simul. Sci. Comput. **7**(1), 1–22 (2016)
9. Epstein, J.M., Axtell, R.: Growing Artificial Societies: Social Science from the Bottom Up. The MIT Press, Cambridge (1996)
10. Forrester, J.W.: Industrial Dynamics. The MIT Press, Cambridge (1961)
11. Gore, R., Lynch, C.J., Kavak, H.: Applying statistical debugging for enhanced trace validation of agent-based models. Simul.: Trans. Soc. Model. Simul. Int. **93**(4), 273–284 (2016)
12. Harris, J.A., Benedict, F.G.: A biometric study of human basal metabolism. Proc. Natl. Acad. Sci. U.S.A. **4**(12), 370–373 (1918)
13. Keim, D.A.: Information visualization and visual data mining. IEEE Trans. Vis. Comput. Graph. **8**(1), 1–8 (2002)

14. Kemper, P., Tepper, C.: Trace based analysis of process interaction models. In: Kuhl, M.E., Steiger, N.M., Armstrong, F.B., Joines, J.A. (eds.) Proceedings of the 2005 Winter Simulation Conference, pp. 427–436. IEEE, Piscataway, NJ (2005)
15. Kornhauser, D., Wilensky, U., Rand, W.: Design guidelines for agent based model visualization. J. Artif. Soc. Soc. Simul. **12**(2), 1–27 (2009)
16. Liu, S., Cui, W., Wu, Y., Liu, M.: A survey on information visualization: recent advances and challenges. Vis. Comput. **30**(12), 1373–1393 (2014)
17. Lynch, C.J.: A Multi-Paradigm Modeling Framework for Modeling and Simulating Problem Situations. Master's thesis, Old Dominion University (2012)
18. Lynch, C.J., Diallo, S.Y.: A taxonomy for classifying terminologies that describe simulations with multiple models. In: Yilmaz, L., Chan, W.K.V., Moon, I., Roeder, T.M.K., Macal, C., Rossetti, M.D. (eds.) Proceedings of the 2015 Winter Simulation Conference, pp. 1621–1632. IEEE, Piscataway, NJ (2015)
19. Lynch, C.J., Padilla, J.J., Diallo, S.Y., Sokolowski, J.A., Banks, C.M.: A multi-paradigm modeling framework for modeling and simulating problem situations. In: Tolk, A., Diallo, S., Ryzhov, I., Yilmaz, L., Buckley, S., Miller, J. (eds.) Proceedings of the 2014 Winter Simulation Conference, pp. 1688–1699. IEEE, Piscataway, NJ (2014)
20. Morland, K., Wing, S., Roux, A.D., Poole, C.: Neighborhood characteristics associated with the location of food stores and food service places. Am. J. Prev. Med. **22**(1), 23–29 (2002)
21. Padilla, J.J., Diallo, S.Y., Lynch, C.J., Gore, R.: Observations on the practice and profession of modelling and simulation: a survey approach. Simul.: Trans. Soc. Model. Simul. Int. 1–28 (2017)
22. Powell, L.M., Chaloupka, F.J., Bao, Y.: The availability of fast-food and full-service restaurants in the united states: associations with neighborhood characteristics. Am. J. Prev. Med. **33**(4), s240–s245 (2007)
23. Rohrer, M.W.: Seeing is believing: the importance of visualization in manufacturing simulation. In: Joines, J.A., Barton, R.R., Kang, K., Fishwick, P.A. (eds.) Proceedings of the 2000 Winter Simulation Conference, pp. 1211–1216. IEEE, Piscataway, NJ (2000)
24. Roza, A.M., Shizgal, H.M.: The harris benedict equation reevaluated: resting energy requirements and the body cell mass. Am. J. Clin. Nutr. **40**(1), 168–182 (1984)
25. Sargent, R.G.: Some subjective validation methods using graphical displays of data. In: Charnes, J.M., Morrice, D.J., Brunner, D.T., Swain, J.J. (eds.) Proceedings of the 1996 Winter Simulation Conference, pp. 345–351. IEEE, Piscataway, NJ (1996)
26. Smith, M.I., Murray-Smith, D.J., Hickman, D.: Verification and validation issues in a generic model of electro-optic sensor systems. J. Def. Model. Simul. **4**(1), 17–27 (2007). https://doi.org/10.1177/875647930700400102
27. Soban, D.S., Salmon, J., Fahringer, P.: A visual analytics framework for strategic airlift decision making. J. Def. Model. Simul. **10**(2), 131–144 (2013). https://doi.org/10.1177/1548512911428987
28. Sokolowski, J.A., Banks, C.M., Diallo, S., Padilla, J.J., Lynch, C.J.: A methodology for engaging modeling and simulation to assess a corollary problem to the obesity epidemic (2012)
29. Sokolowski, J.A., Garcia, H.M.: The significance of modeling and visualization, pp. 33–48. Wiley, New York (2016)
30. Takadama, K., Kawai, T., Koyama, Y.: Micro-and macro-level validation in agent-based simulation: reproduction of human-like behaviors and thinking in a sequential bargaining game. J. Artif. Soc. Soc. Simul. **11**(2), 1–17 (2008)
31. Tolk, A., Diallo, S.Y., Padilla, J.J., Herencia-Zapana, H.: Reference modelling in support of m&s foundations and applications. J. Simul. **7**(2), 69–82 (2013)
32. Tufte, E.R., Graves-Morris, P.R.: The Visual Display of Quantitative Information, vol. 2. Graphics Press, Cheshire (1983)
33. Vernon-Bido, D., Collins, A., Sokolowski, J.: Effective visualization in modeling and simulation. In: Proceedings of the 2015 Spring Simulation Multi-Conference - Annual Simulation Symposium, pp. 33–40. Society for Computer Simulation International, San Diego, CA (2015)

34. Wenzel, S., Bernhard, J., Jessen, U.: Visualization for modeling and simulation: A taxonomy of visualization techniques for simulation in production and logistics. In: Chick, S., Snchez, P.J., Ferrin, D., Morrice, D.J. (eds.) Proceedings of the 2003 Winter Simulation Conference, pp. 729–736. IEEE, Piscataway, NJ (2003)
35. Wilkinson, L., Friendly, M.: The history of the cluster heat map. Am. Stat. **63**(2), 179–184 (2009)
36. Xiang, X., Kennedy, R., Madey, G., Cabaniss, S.: Verification and validation of agent-based scientific simulation models. In: Proceedings of the 2005 Spring Simulation Multi-Conference - Agent-Directed Simulation Symposium, pp. 47–55. Society for Computer Simulation International, San Diego, CA (2005)

Simulating the Ridesharing Economy: The Individual Agent Metro-Washington Area Ridesharing Model (IAMWARM)

Joseph A. E. Shaheen

Abstract The ridesharing economy is experiencing rapid growth and innovation. Companies such as Uber and Lyft are continuing to grow at a considerable pace while providing their platform as an organizing medium for ridesharing services, increasing consumer utility as well as employing thousands in part-time positions. However, many challenges remain in the modeling of ridesharing services, many of which are not currently under wide consideration. In this paper, an agent-based model is developed to simulate a ridesharing service in the Washington, D.C. metropolitan region. The model is used to examine levels of utility gained for both riders (customers) and drivers (service providers) of a generic ridesharing service. A description of the Individual Agent Metro-Washington Area Ridesharing Model (IAMWARM) is provided, as well as a description of a typical simulation run. We investigate the financial gains of drivers for a 24 hour period under two scenarios and two spatial movement behaviors. The two spatial behaviors were random movement and Voronoi movement, which we describe. Both movement behaviors were tested under a stationary run conditions scenario and a variable run conditions scenario. We find that Voronoi movement increased drivers' utility gained but that emergence of this system property was only viable under variable scenario conditions. This result provides two important insights: The first is that driver movement decisions prior to passenger pickup can impact financial gain for the service and drivers, and consequently, rate of successful pickup for riders. The second is that this phenomenon is only evident under experimentation conditions where variability in passenger and driver arrival rates are administered.

1 Introduction

Ridesharing services such as Uber and Lyft have been experiencing explosive growth driven by rider demand and a number of other factors in industry [1]. One challenge

J. A. E. Shaheen (✉)
ORISE Intelligence Community Postdoctoral Fellow, Department of Computational and Data Sciences, George Mason University, 4400 University Drive, MS 6A2, Fairfax, VA 22030-4444, USA
e-mail: jshaheen@gmu.edu

T. Carmichael et al. (eds.), *Complex Adaptive Systems*, Understanding Complex Systems, https://doi.org/10.1007/978-3-030-20309-2_7

for both ride-sharing service providers as well as for their driver contractors is how to maximize driver acceptance of new customers while ensuring drivers gain maximum utility from rides given. In other words, how to maximize both financial gains for the service and its drivers while ensuring maximum service-level quality for its customers.

Perhaps as a reflection of the growth of the ridesharing industry and of the aforementioned challenges, empirical research in this area is also experiencing a surge, exemplified by a growing number of journal publications [1–18] that explore the multidimensional challenges and opportunities produced by the widespread adoption of ridesharing services.

This paper aims to investigate spatial behavioral conditions under which drivers can gain increased financial returns (utility) on their invested time, while simultaneously ensuring that a maximum number of potential passengers reach their destination. The model produced utilizes an agent-based modeling (ABM) framework that has the potential to be extended, expanded, and tested under many variable conditions. And therefore, while it would be immediately salient that much can and should be tested with the model, we reserve future extensions and testing for future papers.

It is worthwhile to note that the agent simulation perspective is highly suitable for testing spatial behaviors; when utilizing agent simulations it is considered trivial to create many autonomous, heterogeneous agents following one or more behavioral rule-sets and to simulate behaviors for various initial conditions and parameters without the constraints of rigid assumptions. Additionally, as will be shown by the results of our experimentation, other modeling techniques may not fully capture the true temporal dynamics of a ridesharing service because of heterogeneity in agent decision-making, the spatial significance to end results, and the variable scenario conditions under which emergent properties could arise. In a subsequent section, we will show the relevance of the later.

The model described in this paper focuses on simulating drivers and riders in the Washington, D.C. metro region and attempts to simulate the movement of drivers under two spatial movement conditions.

Ultimately, the aim of this model is to gain insight into whether drivers, riders and ridesharing services benefit more or less from optimized decision-making during the drive-pickup-drop-off lifecycle familiar to ridesharing customers, while the aim of this paper is to highlight testing of some specific conditions.[1] In later iterations of the model, an increased variety of behaviors will be investigated.

1.1 Background

There have been a number of studies in the last few years taking aim at understanding ride-sharing services and carpooling schemes—each of which takes a different

[1]The author of this paper registered with one ridesharing service in order to gain insight into the natural behaviors of drivers and riders of the service. A total of 30 trips were carried out.

investigative position on the challenges faced by resource pooling services as a whole [1, 22], while some consider some facet of modeling behaviors [3] using agent-based approaches. The majority of papers reviewed were of modeling carpooling decisions as an optimization problem [7, 11, 12] and finally, some approaches intended to make early-stage predictions about carpooling and ridesharing trends [6, 9] were pre-existent in the literature.

What becomes very clear during a topical literature review over the last few years is that no real attempt had been made to provide for a comprehensive ridesharing agent-based simulation that captures prevalent dynamics; though much of the research attempts to understand the effects of ridesharing in general.

For example, Cho et al. [3] provided a full description of a hypothetical agent-based model for a carpooling application without offering an actual build of the model hypothesized. The authors focused on the systemic theoretical structure of the proposed model, the mathematics and optimization techniques that *would* be used and the general form of social network types that *could* be used between the driver agents of said model. The same group [4] later proposed another agent-based model—this time only based on social network interactions without implementation.

Significant advances in the area of heuristics and algorithm development that propose better route optimization techniques have also been made over the last few years and this is an area where high-value and productive work has been put forward. For example, Pelzer et al. [12] developed a method which aims to best utilize "ridesharing potential while keeping detours below a specific limit" using a spatial partitioning method.

IAMWARM aims to build a foundational baseline to test a small number of interesting spatial problems for which answers have not been provided as of yet and to use the model created as the basis for future improvements, extensions, expansions, and experiments. We begin that endeavor by discussing our primary and most central question: Can ridesharing utility for both riders, drivers, and service be increased through varying the information-shared among agents, ultimately affecting spatial movement behaviors? And if so, under what conditions could one note a difference in system-level properties?

1.2 Information Asymmetry Versus Information Symmetry

Our path is to find the simplest method of testing ridesharing utility schemes spatially, so we begin by discussing the problem of increased information sharing briefly implied in the previous section.

For our specific context, we define information-asymmetry as a lack of information regarding the location of other drivers by other drivers. That is—drivers, in an information-asymmetric service, would not be given the locations of other drivers, or riders,[2] with exception of a single potential rider within their vision's radius who has just requested a pick-up, and thus without that information and without clear route planning driven by spatial demographics, drivers would simply move about randomly hoping to 'luck out' and be near a potential customer when they request a pickup. This is currently the method by which all ridesharing services manage their respective platforms. Drivers of those services are not given location information of other drivers, and must move about based on randomness, their own past experiences and information gleaned from their social networks; and so ultimately, must make their spatial movement and positioning decisions based on either luck or experience gained from learning. We will omit learning behavior from this iteration of the model.

Symmetry represents a condition such that driver agents have all the available information about other driver agents and rider agents. For this model however we bound true information symmetry to a localized version that limits driver agents' knowledge to the nearest driver agent and only to the nearest rider agent. The comparison between information asymmetry and symmetry which will be established by the comparison between random movement and Voronoi movement will be applied such that agents have no vision for information asymmetry scenario runs (no knowledge of the position of any other driver) and have only local vision in the information symmetry variation of the model (knowledge of the nearest driver agent's position). This modification is a direct result of the platform chosen for the development of the simulation and its ability to perform, and due to a lack of a clear theoretical or even observed cognitive standard to base spatial behaviors upon in this case.

This model will investigate how drivers would benefit from having local information about peer drivers available in real time using a hypothesized spatial behavior and whether that additional information would maximize driver utility. And, though there are a number of differing spatial behaviors that can be considered, we will only test one behavior which we theorize would result from access to that information. We will extend this model in future iterations with more behaviors.

We call the assumed behavior resulting from information symmetry *Voronoi* behavior or *Voronoi* movement. We propose this behavior using the spatial concept of a Voronoi polygon [19] as a base. A Voronoi polygon or diagram—as it is commonly known—is a partitioning of a spatial plane such that "all locations in the Voronoi polygon are closer to the generator point of that polygon than any other generator point…in Euclidian plane" [18]. In other words, it is the space such that maximum territory is created for each generator point without overlapping the area belonging to any other generator point. Figure 1 shows an illustration of generator points and their respective Voronoi polygons.

Voronoi movement essentially amounts to driver agents receiving location information about the nearest driver agent and moving away from them so as to increase the potential of picking up a new customer and reducing local competition—a

[2] We will later explain our terminology in detail, but for now we define a driver as an agent who is picking up a rider from one location on our model's spatial grid to another. Once a rider is "picked up" we will refer to him as a passenger. In our model a passenger is no longer an agent but is a data point in the driver agent's attributes list.

Fig. 1 Shows an example of a Euclidean space with generator points and their corresponding Voronoi polygons. The polygons are "emerged" from the collective positions of the generator points such that all points in each area corresponding to a polygon are closer to the corresponding generator point than to any other point [20]

diverging topological behavior where each agent maximizes the distance and the territory between self and all other agents. This viewpoint is a corollary to the Voronoi polygon—from the view of the generator point (agent) not the adjacent spatial points in the polygon, hence we call this behavior Voronoi behavior.

We compare Voronoi movement behavior with a random movement pattern where driver agents move randomly across our spatial grid until they are close enough to a rider agent to execute a pickup. The random movement behavior is a reflection of driver agents having no knowledge of where other drivers and where customers might be. In this iteration of the model, we assume agents do not learn.

In a high-fidelity model, it would likely be the case that drivers would learn about rider behaviors and adjust their own behaviors accordingly. However, as you will see in the results section, even with this simplification, our baseline model offers interesting conclusions nonetheless. Finally, rider (potential passenger) agents do not move in this iteration of the model but enter the simulation at a spatially random location on the Washington, D.C. geographic lattice. Figures 2 and 3 show the graphical representation of the model which was developed in NetLogo [20] and subsequent sections will discuss model design particulars.

2 Model and Methods

IAMWARM was implemented in NetLogo 5.3 and utilized the GIS extension native to the platform to import map and GIS data into the model. The model's spatial configuration was based on a road network imported from the Washington, D.C.

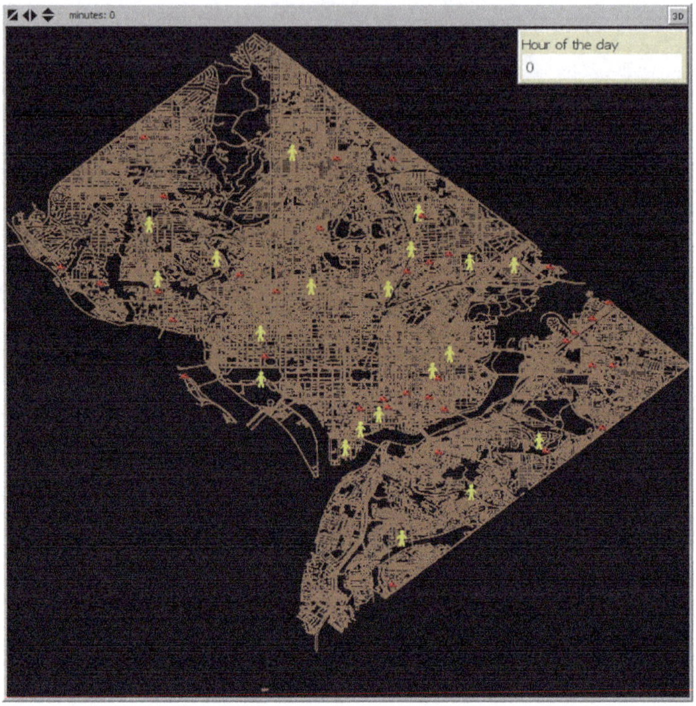

Fig. 2 This is the graphical representation of the model. Yellow agent types (person icons) are rider agents. Red (car icons) are driver agents. When a driver agent performs a pickup, their color turns from red to white to display that they are no longer available

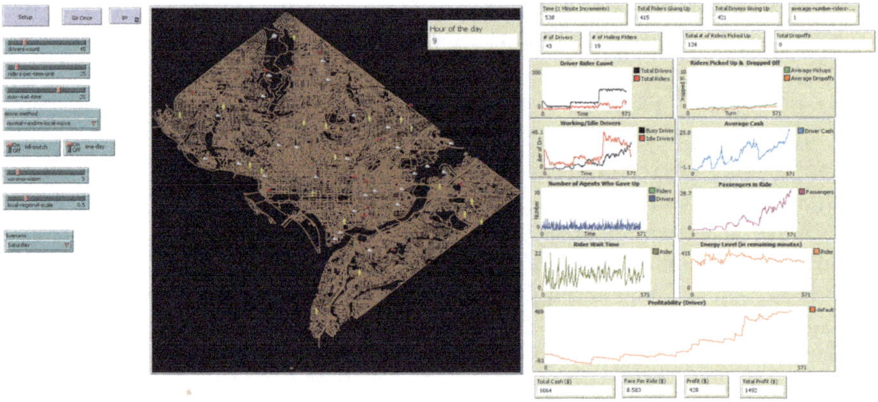

Fig. 3 This model was implemented in NetLogo 5.3 and utilized open access data from the Washington, D.C. government website. In this figure, we show the graphical user interface of the model. Inputs, such as the number of drivers and the number of potential riders active at any moment are complemented by outputs on the far right such as profitability, average cash on hand, and passenger pick-ups

government Open Data Project website,[3] which included highly accurate, editable shapefiles. The data included feature labels for roads and intersections. These features are used in the instantiation and location initialization of agents, and for certain critical agent behaviors to be discussed in later sections of this paper.

At the current iteration of the model we chose not to include additional layers of geographic information for simplicity (only the road network was included), but in future iterations utilizing the Open Data project more broadly can be advantageous in increasing the efficacy of our model specifically by adding more spatial configuration data. Figure 4 provides a complete graphical summary of the model's logic.

The road network was imported to NetLogo and an internal spatially-equivalent configuration was assigned (labeling). All roads were labeled internally by a variable to help identify spatial cells that contained a road, versus spatial cells that did not. This would later be an important step when designing the movement choices of agents in the simulation since all movement and agent entry will occur on road cells as one might expect.

The model contains two agent types: drivers and riders. Drivers can move across the model space, but only on cells that contain a road, while riders do not move, but can only be initialized on roads, specifically intersections.

The movement of the driver agents was designed to be based on a direct line of sight—that is—although the drivers must always remain on roads, we assume that following actual traffic routes would not provide a negligible difference in destination arrival times. This is mainly due to the size of our spatial lattice which numbers in the several thousand. Moreover, for our research goals, it suffices that driver agents move in a direct path to their destinations once rider agents are picked up. In future iterations of the model, traffic and road direction movement could be taken into account to create a greater sense of realism. For now, driver agents move on roads in a direct fashion to their destinations.

2.1 Agent Behaviors

Before we discuss the specifics of agents used in our model we define the terminology used in the model. We define driver agents as those agents who are intending to pick up a rider. A rider agent is an agent who has been instantiated and can be picked up by a driver. Once a driver agent picks up a rider agent, the rider agent becomes a passenger. Passengers are not agents and do not interact with their environment. In other words, riders who become passengers simply become an attribute of the driver agents, releasing with them certain data points which are then used by the driver agents post pick-up. This terminology will be used throughout the model description.

---

[3]www.dcogc.org.

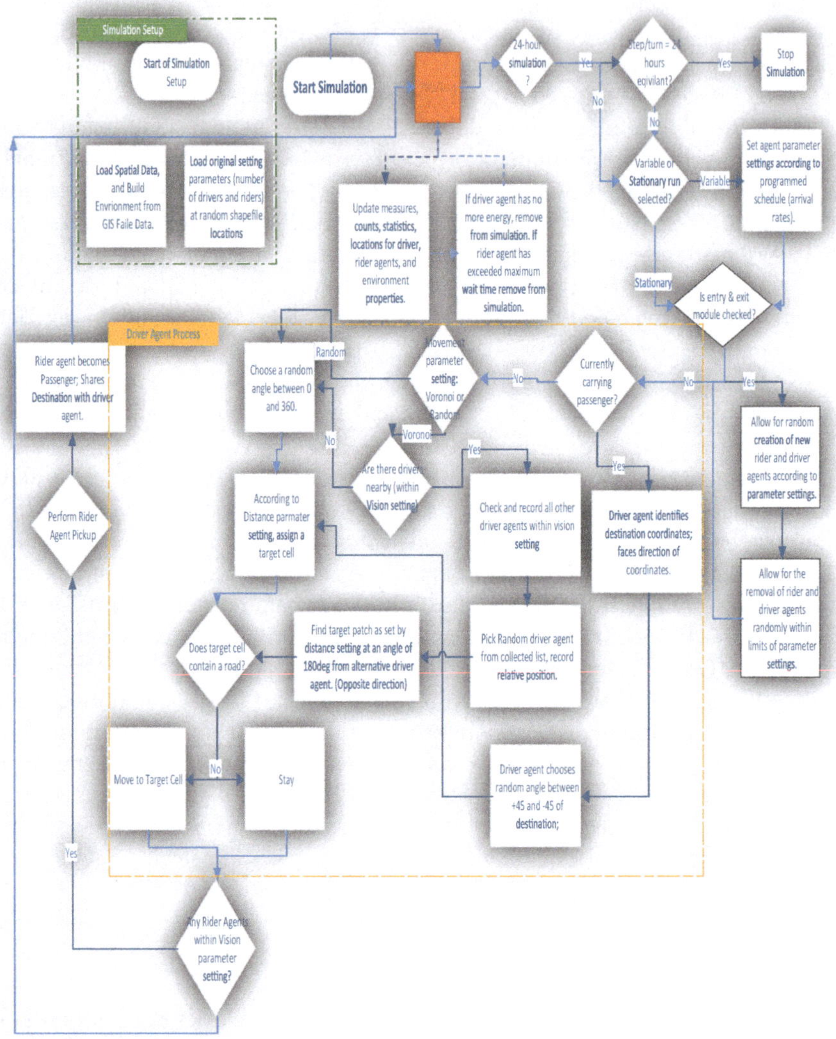

Fig. 4 A complete breakdown of the model's process is presented in this figure. The area labeled inside the yellow box is where the majority of driver agent behavior takes place. The small orange box at the top of the diagram represents the model moving to the next time unit/time cycle. Note that the majority of rider agent interactions are included as part of the driver agent behaviors since much of the rider agent behaviors in the model is restricted to entering or leaving the simulation

2.2 Driver Agents

Driver agents are instantiated in the initial setup of the model according to a user input parameter and a random distribution set to uniform properties. Spatially, the driver agents are initialized at random road intersections of the geospatial model, and according to a randomization test score that allows for a greater randomized spatial distribution of agent objects. Driver agents are also instantiated throughout the model run according to the aforementioned user set parameter. Those agents are also instantiated at road intersections and their rates of arrival are also set according to a user set parameter.

Driver agents move from one spatial cell to another by first checking whether a road exists in the cell ahead. If it does, then an agent may move to that cell. If no road exists in that cell then for the random movement behavior without an active passenger procedure, agents are instructed to rotate a random number of degrees between 0 and 360 and repeat the process.

For driver agents who are carrying a passenger, the process is the same. However, the randomized degree value is set to be between −45 and +45 degrees if a road is not found directly in the heading of the driver agent. By doing so, we ensure that drivers are constantly moving in the direction of their destination, but are still able to overcome the majority of obstacles in their way, such as the lack of available roads to travel on. This method does have grounds in reality in that drivers who may not necessarily know precisely how to optimize their routes, may in general, pick a random route that they know to be in the general direction of their destination. In future iterations of the model a more advanced pathfinding algorithm could be adopted such as the A* pathfinding algorithm, but for our purposes, we assume that the difference is negligible and it is trivial to show so through a model run.[4]

Driver agents are assigned a number of attributes at instantiation and some are assigned as the model is run situationally. Attributes include energy level, cash-on-hand, time driven, current driver destination (if carrying a passenger), riders who are nearby, current passenger id, time the current passenger has been on a trip, how many riders the driver has picked up, and how many passengers the driver has dropped off, as well as a Boolean passenger variable indicating if the driver agent is currently carrying a passenger. The attributes are more critical to the progress of the model at varying times through simulation runs, depending on the active phase of the drive-pickup-drop-off-drive cycle of the driver agents. Therefore, a deeper explanation of the attributes and their relevance is appropriate at this time.

We assign a level of energy to every driver agent set to be a random number following a normal distribution between a 4 and 8 h range translated into model time units (which is set to be at 1 min per time unit) by estimation. The underlying assumption is that the majority of drivers will only be able to drive, regardless of their level of success, for a period determined by time availability and physical and/or mental fatigue. This is an appropriate assumption and the Gaussian shape of this distribution

[4]For a video of a typical model run, please visit https://www.youtube.com/watch?v=apJEvDl4aqc.

is not far-fetched. Time-driven is a variable that counts the amount of time driven by each driver and is used to display and calculate the model's summary statistics.

Cash-on-hand is the variable attribute that stores the accumulation of fares each driver agent has gained, as well as the variable in which cash is deducted (vehicle and transportation costs) for drivers who are not carrying active passengers. In other words, it is the driver agents' total utility and financial gain at any given time-period.

Pick-up count and drop-off count are variables that store the total number of riders successfully picked up from their initialized location and successfully dropped off at their destination, respectively. The Boolean passenger variable shows whether the current driver is currently seeking a rider or already has a passenger (as discussed earlier, riders that receive a pickup by a driver agent become 'passengers'—simply an attribute of the driver agent) and is used in a number of important model mechanics. Passenger-id is a variable that stores the id of the rider currently within the driver agent's vehicle when she becomes a passenger. It is equivalent to the driver asking for the rider's name prior to pick-up and is used in the verification process of the model to ensure that drivers are successfully picking up intended riders.

Variables for nearby riders and nearby drivers were also implemented as attributes of the driver agents. The first being the number, and id of any nearby riders waiting for pickup. This attribute is used to evaluate whether there are any riders nearby available for pickup. The second is whether there are any nearby driver agents, and is used in the Voronoi movement mechanism. The details of the Voronoi movement mechanism will be discussed in a subsequent section.

2.3 Rider Agents

Rider agents are instantiated at initialization of the model at random locations (intersections). Rider agents do not move but await their intended pick up in the same location. This is in line with expected behaviors of ridesharing service customers. Rider agents are instantiated utilizing a user-set input and a random variable to allow for some stochasticity in the model runs. Spatially, and in a similar fashion to driver agents, they are placed randomly across the available intersections of the model's geographic configuration, and only on road cells.

Rider agents are instantiated throughout the model's runs but at a rate per time unit user fixed parameter, unlike driver agents who typically remain in the model's space until they decide to *leave* (randomly) or because they have exhausted their energy variable and are replaced stochastically up to a maximum user set parameter. In other words, while the number of driver agents is set by a maximum capacity global (exogenous) variable, the number of rider agents is set to be a rate of arrival following a probability distribution. For simplicity, we chose the arrival rate probability distribution to be normal, though in subsequent iterations of the model testing of other probability distribution types, such an exponential arrival function, would be necessary.

Rider agents possess two attributes for this iteration of the model. The first is the rider destination, which is a randomly assigned destination converted to the spatial coordinate equivalent. This destination is assigned at the moment of instantiation of the rider agents. The second is the wait time variable which is a count of how long a rider has been waiting for pickup and which is used to compare to a user-set input to determine whether the rider agent should look for 'alternative' transportation methods (like metro or bus service). A user designated input allows for varying the waiting time of rider agents. For typical model runs we assigned this variable to 20 min. Once a rider agent reaches their maximum waiting time assigned they leave the simulation. This is a proxy behavior for the rider agent attempting to find alternative modes of transportation to their destination.

2.4 Model Mechanics

The model relies on user inputs for the number of drivers (capacity), the number of riders arriving per given time unit, the maximum waiting time for rider agents, whether or not to use random movement while attempting to find a rider or whether to maximize distance from any other drivers in the area of assigned vision (Voronoi movement). Based on these inputs and the parameters of the model a typical model run behaves as follows.

Rider and driver agents are instantiated on a highly accurate road map of Washington, D.C. at random intersections in accordance with the input parameters assigned by the model user. Once the model is run, rider agents are spawned while others, according to the user input, will leave the simulation. The same is applied to driver agents.

Driver agents move according to one of the predetermined movement methods (random movement or Voronoi movement based on which version of the simulation is run) until they are within a user-set proximity setting of a rider agent (for our model the vision was set to 3 cells). During this time, driver agents lose energy at a rate of 0.75 per time unit and lose cash at a rate of \$0.1 per time unit.[5]

Once a driver agent is within a designated (by modeler) vicinity of the rider agent, a pick-up occurs, and the rider is converted to a passenger. A transfer of the destination of the rider occurs from rider agent to the driver agent at this time. Once a driver agent has successfully executed a pickup, the agent can no longer make any additional pickups, and it is at this point that the driver agent begins to accumulate 'cash', set at a rate of *\$2.00 + 0.60/*time unit. Calibration of the model was conducted to reach a dollar amount that could be probable through qualitative observations of distance and time of real-world trips versus the model's spatial geometry. The key was to set the fare rate to include a fixed amount and a variable amount so as to reflect actual ridesharing services.

[5]We assigned these cash variables based on a rough estimate of distance travelled versus fare/ride gained from observations and experiences with a ridesharing service.

Driver agents then proceed in a direct path to the coordinates of the transferred destination while earning ‘cash’ and losing ‘energy’. Once they reach their proximate destination, a drop-off is executed, their passenger-carry variable is reduced from 1 to 0, and a successful trip is recorded as being now completed. All relevant attribute and model-level variables are updated with this new information. The driver then continues to move searching for new riders and repeats the drive-pickup-drop-off process.

Rider agents who are not picked up within their waiting time-period limit find alternative transportation and leave the simulation, while driver agents who are not carrying a passenger could “give up” and leave the simulation. The latter could also run out of energy and leave the simulation due to fatigue. Typical runs are for a 24 h period, but a model user can run the model indefinitely if they desire.

2.5 Model Inputs and Parameters

We implemented 7 inputs in our model that can be assigned and varied by the user. Table 1 summarizes those inputs and contains their descriptions. The most important of which are the maximum capacity for driver agents (drivers-count) and the rate of entry of new rider agents into the model (riders-per-time-unit). Other inputs are also critical but were not tested in a significant way—though adjusted for calibration and realism. Those are the Voronoi vision setting which controls how far driver agents can see other driver agents, the local-regional scale which amounts to an adjustment for the speed of movement of the driver agents, and a binary-switch which turns on or off the possibility for both driver and rider agents leaving the simulation randomly.

In Table 2, we describe the parameters and distributions used in various parts of the model. As mentioned in an earlier section we used a normal distribution of varying means and standard deviations as the basis for a number of statistical tests and parameter values so as not to add any additional unverified assumptions or complexity to the model. Hence, we rely on the Central Limit Theorem heavily. However, in future iterations of this model specific testing and data collection of the distributions’ parameters must be undertaken and compared to real data from a ridesharing service.

2.6 Model Outputs

A number of outputs were included in the model to assist in the verification process, to understand model mechanics and to derive results from model runs. Table 3 lists those outputs and their descriptions. Our focus was to understand driver agent utility given some set of inputs, parameters, and pre-conditions. There are many forms of driver agent utility to consider, each of which would require a focus on a different set of output measures. For this iteration of the model, we chose to focus our attention on total driver agent utility in the form of total profit from each model run. We

Table 1 Inputs of the model with typical value assignments

Type	Input	Description	Typical value(s)
Driver agents	Driver agent number	Assigns a maximum number of drivers to be active at any given time unit	50–150
	Normal random movement (choice)	Sets the movement behavior of drivers to be of a random nature while they await a rider pickup	N/A
	Voronoi movement (choice)	Sets the movement behavior of driver agents to follow a Voronoi-distance maximizing method	N/A
	Voronoi vision	If Voronoi movement is chosen, sets the Voronoi movement vision distance	3
	Local-regional scale	Sets the vision and movement range for driver agents. This amounts to a speed setting and is used to calibrate the model	0.5
Rider agents	Riders active per time unit	Sets the rate by which new riders enter the simulation and await pickup	20–75
Environment	Scenario (choice)	Sets the model into a run type where an expected rate of arrival for riders and an expected maximum capacity for drivers is set at different hours of the day	Saturday

include no outputs to measure ridesharing service utility or rider/passenger utility in our final analysis and conclusions, however, a number of outputs aimed towards the measurement of rider agents, passengers, and ridesharing service utility are designed into our model and are displayed to the user. We hope to expand on our analysis of system utility by considering rider, passenger and service utility in future iterations of the model.

2.7 Scenarios

Much of the model can be run in a stationary mode—that is—it can be run in a form of equilibrium where driver agents and rider agents arrive at predetermined rates resulting in a constantly changing but variably fixed dynamic. This is interesting for general runs, verification of model mechanics and quality, as well as to gain a

Table 2 Parameters used in model mechanics

Parameter	Type	Value	Description
Driver agent placement	Test X > 0.5	Normal (1,1)	Tests whether a random number from a normal distribution with a mean of 1 and standard deviation of 1 is greater than 0.5. If so, placement of a driver agent succeeds at a given intersection
Rider agent placement	Test X > 0.5	Normal (1,1)	Tests whether a random number from a normal distribution with a mean of 1 and standard deviation of 1 is greater than 0.5. If so, placement of a rider agent succeeds at a given intersection
Driver agent energy	Attribute	Normal (360,120)	Sets the energy of a driver agent at instantiation as a number drawn from a random distribution with a mean of 360 and a standard deviation of 120 (min)
Kill count	Variable	|(Normal (0,1)|	Sets the number of driver and rider agents who will leave the simulation, randomly without depleting their energy (driver agents) or reaching maximum wait time (rider agents) to be the absolute value of a random number drawn from a normal distribution with mean 0 and standard deviation of 1

general understanding of the ridesharing process. It is trivial to hypothesize that in any transportation system the rates by which riders and drivers arrive, interact, and exit are variable but also subservient to the city (the spatial lattice) in which the ridesharing service operates. This would include seasonal variables such as the time of year, month, day, and time of day. Additionally, rider and driver rates and activity are also affected by current events, traffic, roadworks, weather patterns and other exogenous factors. Therefore, though running the model in situ yields important insights, it is important to run experimentation in some variable scenario for comparative reasons and for a closer approximation of real-world dynamics simply because the variance itself could yield insight.

Table 3 Model outputs and measures

Output	Description
Number of driver agents active	The number of driver agents active in the model
Number of rider agents active	The number of rider agents active in the model
Total riders giving up	The total riders giving up based on randomly set parameters
Average number of riders picked up per time unit	The average number of riders picked up per time is calculated for each time unit and displayed
Total number of rider agents picked up	This is the total number of rider agents converted to passengers
Total number of successful drop-offs	Total number of successful drop-offs, which tends to be lower than the number of pickups as some driver agents don't reach their destinations
Number of idle driver agents	Number of driver agents without an active passenger
Number of working driver agents	Number of driver agents with an active passenger
Average cash on hand	The total amount of cash held by all driver agents (active)
Number of agents who left (randomly)	Number of agents who left the simulation due to random tests
Number of passengers in active trips	Number of passengers carried by driver agents
Average wait time	Rider agent average wait time
Average energy level	Driver agent average energy level
Total cash with active driver agents	Total cash for all active driver agents at any given time unit
Average fare per ride	Average fare per ride at any given time unit
Total profit generated	Total profit generated by all activity of the model

Therefore, based on anecdotal evidence gained from the author's registering with a ridesharing service and gaining first-hand experience in typical driver decisions made, we develop a scenario which is not entirely hypothetical in order to test the model's effectiveness under varying conditions. In future iterations of this model, we intend to develop scenarios grounded in real data collected and to develop a number of them to test different scenarios under different conditions without such heavy reliance on qualitative observations.

For this model iteration, we conducted a test of one scenario—the "Saturday" scenario which varies only arrival rates of both driver and rider agents according to what might be expected on a typical weekend day—Saturday. Table 4 provides a summary of the scenario and reasoning, where appropriate, for selection of scenario

Table 4 This is a variable model scenario used in testing the model under realistic conditions drawn from anecdotal observations

Hour	Driver capacity	Rider rate of entry	Reasoning/explanation
5 AM	10	5	Early morning—airport traffic—mostly quiet
8 AM	20	10	
11 AM	50	20	Drivers starting their day for the Saturday brunch/lunch
12 PM	5	10	Lunch time—low activity
1.30 PM	45	25	Post-lunch rush
2.30 PM	50	15	Post-lunch rush
3.30 PM	25	10	Stationary activity
5.30 PM	40	5	Evening drivers beginning their shifts
6.30 PM	45	5	Evening drivers beginning their shifts
7.30 PM	60	30	Night activity period—riders are going out to social events
8.30 PM	80	40	Night activity period—riders are going out to social events
9.30 PM	100	40	Night activity period—riders are going out to social events
10.30 PM	90	10	Low rider activity—riders are at their destinations. Drivers still on the road expecting a rush of new riders
11.30 PM	80	10	Some drivers give up, exit the simulation
12.30 AM	75	30	More drivers give up. Riders beginning to end their work shifts
2 AM	65	30	More drivers give up. Riders beginning to end their work shifts
4 AM	35	10	End of night traffic. End of 24 h cycle

inputs and parameters. As you will see from the table, we varied the arrival rates and expected capacities of rider agents and driver agents respectively. For example, on 'Saturday' we would expect high customer demand for the hour before 'lunch' as many city dwellers may be engaging in social activities in the subsequent hour and so they intend to arrive before 'lunch-hour'. In this time-period (11 AM) and in this specific scenario, we expect that in anticipation of higher customer demand the number of driver agents may also increase, and so we increase the maximum capacity of the driver agents.

Table 5 Verification and validation methods

Goal	Method	Result
Verify that road network imported correctly	Display and check spatial cell attributes	Success
Verify that agents instantiate on roads, specifically intersections	Compared spatial coordinates of agents with cell coordinates of intended intersection	Success
Verify that agents instantiate with the correct attribute values	Displayed agent attributes at random	Success
Verify that the movement of agents is as intended	Visual observation and numerous model runs	Success
Verify that the model mechanics for rider entry functions as intended	Raised and lowered the rider entry value and monitored expected increases or decreases in outputs	Success
Verify that the rider capacity functions as intended	Raised and lowered the rider entry value and monitored expected increases or decreases in outputs	Success
Validate pickup mechanics	Verified through the transfer of rider id and destination	Success
Validate drop-off mechanics	Observed attribute changes for increases in drop-off values	Success
Validate agent attribute changes through model runs	Through many runs and observations	Success

2.8 *Testing, Verification, and Validation*

To test whether spatial movement behaviors can affect driver financial gain, we aimed to compare Voronoi movement prior to agent-pickup with random movement prior to agent-pickup. In other words, having drivers access information about other local drivers and having chosen to 'spread out' maximizing their personal territory and likelihood of rider pickup, when compared with random movement, regardless of where other driver agents may be. This collection of tests translates to:

1. Comparing Voronoi movement with random movement under stationary conditions (constant capacity and arrival times through entire run).
2. Comparing Voronoi movement with random movement under variable conditions, namely a 'Saturday' scenario (varying arrival rates and capacity for rider agents' entry and driver agent entry).

Therefore, to test our model we conducted 4 standard runs: A scenario-based set of runs with a comparison of random movement and Voronoi movement, and a stationary standard run with Voronoi movement and random movement. We also conducted a number of verification and validation tests to ensure that the model is run correctly as well as that it is running as intended. We summarize those efforts in Table 5.

3 Results

3.1 A Typical Run

We present typical model results for the 4 possible variations of our test runs. Figures 5, 6 and 7 and Tables 6 and 7 show plots and summary results of our model run. Figures 5, 6 and 7 contain the total and accumulated financial gain of all agents with total profit on the y-axis and time units on the x-axis. This measure includes the financial gain made by agents who have left the simulation due to fatigue or for any other reason.

We see no real and substantial difference in terms of total profit (financial gain) between either of our stationary conditions model runs for random movement and Voronoi movement (yellow and grey). That is—whether drivers chose to 'spread out' or move about randomly in the hopes of picking up more customers did not affect, on average, their financial gain. In fact, the difference was comfortably within 2 standard deviations for both Voronoi and random movement runs.

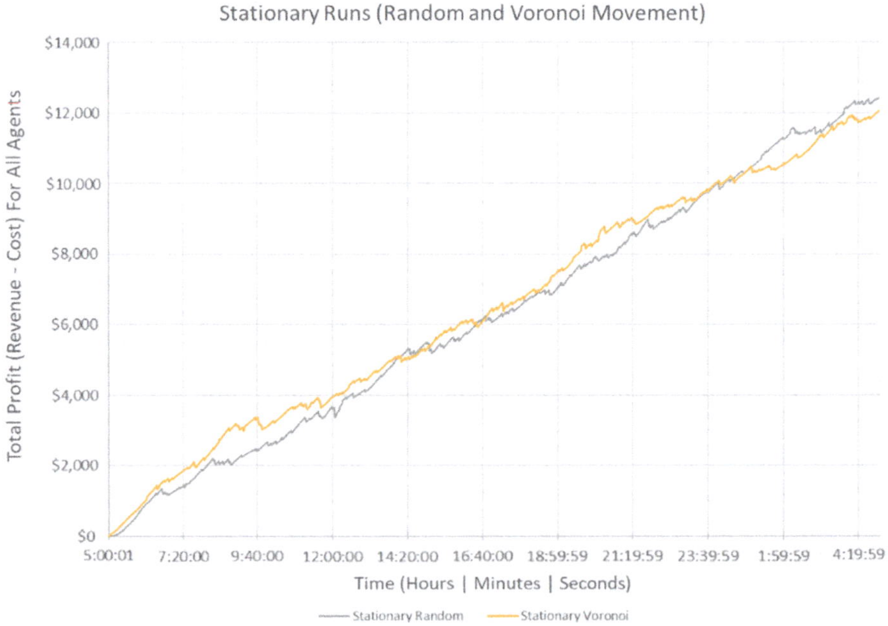

Fig. 5 This figure shows the total profit made by all driver agents over a 24 h period—a single run of the model for both random driver movement (gray) and Voronoi driver movement (yellow). The figure shows that while there are stochastic gains made at different times in the model run by both random and Voronoi movement behavior, there are no clear advantages in utilizing either behavior type when the arrival conditions of drivers and riders are stationary

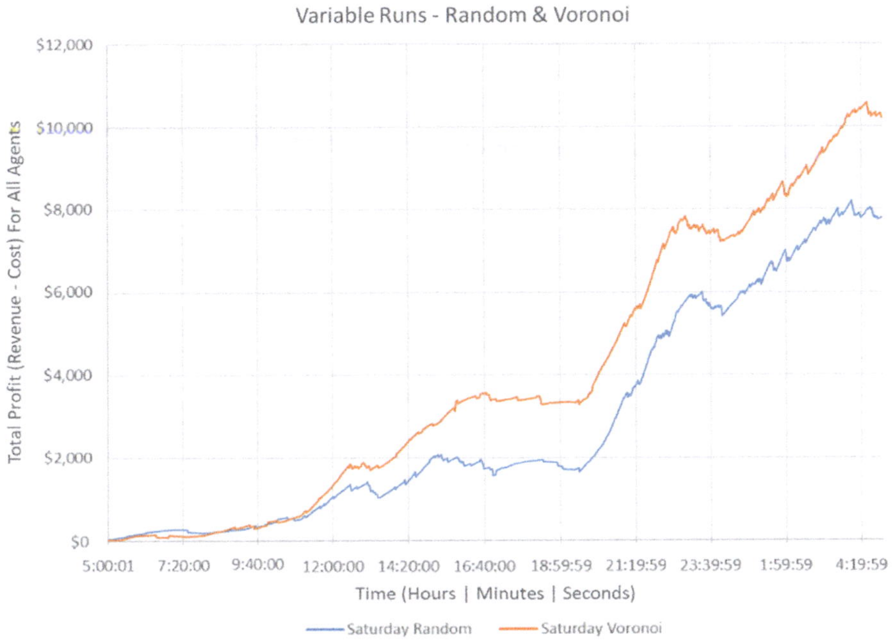

Fig. 6 This figure shows the total profit made by all agents over time. Not the divergence between Voronoi movement (Orange), and the random movement (Blue) under variable conditions. Voronoi movement outperforms random movement

Table 6 Summary statistics for stationary run for both random and Voronoi movement

Stationary random		Stationary Voronoi	
Mean	$ 6,284.02	Mean	$ 6,483.38
Standard error	$ 94.35	Standard error	$ 88.71
Median	$ 6,161.00	Median	$ 6,400.90
Standard deviation	$ 3,580.49	Standard deviation	$ 3,366.40
Maximum	$ 12,426.55	Maximum	$12,059.45

Table 7 Summary statistics for variable properties run for random and Voronoi movement

Variable (Saturday) random		Variable (Saturday) Voronoi	
Mean	$ 2,845.05	Mean	$ 3,931.65
Standard error	$ 68.96	Standard error	$ 87.31
Median	$ 1,828.90	Median	$ 3,332.50
Standard deviation	$ 2,617.00	Standard deviation	$ 3,313.05
Maximum	$ 8,195.65	Maximum	$ 10,565.45

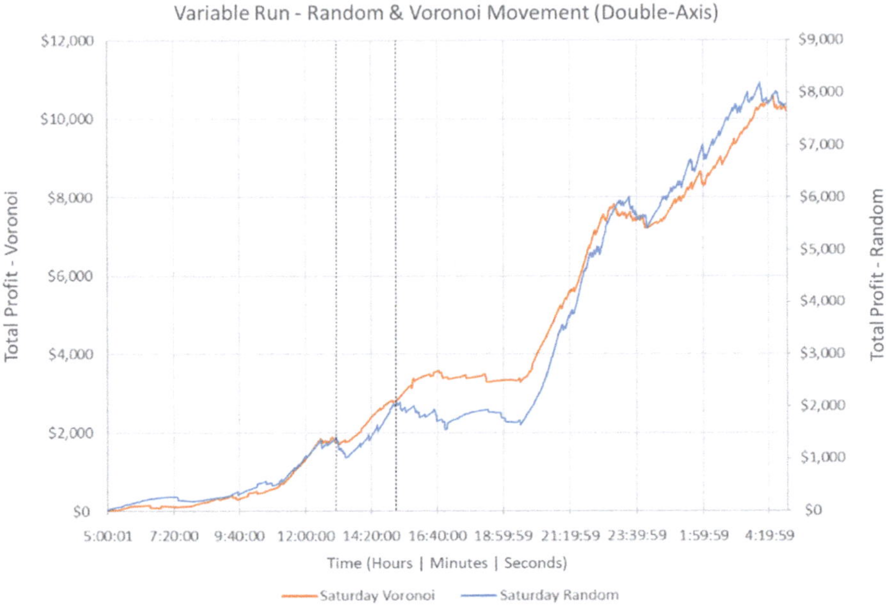

Fig. 7 This figure utilizes 2 axes, one for the variable conditions scenario run (Saturday) Voronoi movement (left axis) and another for the random movement (right axis). This makes for easier visual comparison and yields insight into precisely which point in time the divergence between the two behaviors would typically occur. In this particular run it is clear that Voronoi movement begins to outperform random movement at 630 min

Remarkably, for the variable scenario runs a stark difference emerged between the two spatial movement types, unexpectedly. Divergence in the profitability between random movement choices and Voronoi movement choices for driver agents was clear, and exhibited in both the total profit made by driver agents in a 24 h run (Figs. 6 and 7) and in the summary statistics of the model run as a whole (Table 7).

Specifically, we can comfortably note that Voronoi movement for driver agents provides greater utility (financial gains) for drivers when varying rates are executed on the agents' arrival rates i.e. when a scenario is utilized. Where for stationary model runs neither movement method prior to rider pickup provided any visible change in driver agent utility. Consequently, our observations and analysis of the model run took a focus on the variable scenario runs, and more precisely on the moments of divergence of the variable scenario random movement run when compared to the Voronoi movement variable scenario model run.

More importantly than the observation that Voronoi movement outperformed random movement only in variable run conditions, is whether we can deduce precisely where the divergence between the two behaviors began to take shape under variable entry conditions.

The first instance of divergence in a typical run occurs at around 1 PM (marked on Fig. 7) into the model run which we hypothesize as being representative of

Table 8 This table shows the relevant arrival rates for drivers (left) and riders(right) for the first point of divergence in the variable conditions model run

11 AM	50	20	Drivers starting their day for the Saturday brunch/lunch
12 PM	5	10	Lunch time—low activity
1.30 PM	45	25	Post-lunch rush

Table 9 This table shows the relevant arrival rates for drivers and riders (left, right) for the second point of divergence in performance of the variable conditions model run

2.30 PM	50	15	Post-lunch rush
3.30 PM	25	10	Stationary activity
5.30 PM	40	5	Evening drivers beginning their shifts

"lunchtime" activity and model as being a constant driver agent capacity of 5 drivers and an arrival rate of new riders of 10. Both rates are a reduction from the 11 AM hour which had a maximum capacity of 50 driver agents and 20 respectively. At 1.30 PM the capacity for new driver entry increased to 45 and the rate of arrival of rider agents also increases to 25 (Table 8). Figure 7 has both the random movement and Voronoi movement drawn with separate y-axis on the same time-scale (x-axis) so as to allow us a better visual comparison of both run-types, and we can see that at this lunch-time hour a slight divergence of performance begins to emerge, allowing drivers who are using Voronoi movement to make placement decisions that outperform those that move randomly.

The most salient divergence between the performance of the two behaviors we tested occurred at around the 3.30 PM time-period. Figure 7 shows the stark difference in performance and thus in utility-gain between the two behaviors. Our scenario at this time-period calls for the decrease of both driver capacity and rider entry from the 2.30 PM period (from 50, 15 to 25, 10, for driver and rider agents respectively). Table 9 summarizes the relevant part of the scenario run.

Thus, we saw a divergence in performance for a case when both driver and rider entry were increasing simultaneously, and for a case where they were decreasing simultaneously as well, which dispels any notion that divergence in performance would solely be due to a decreasing rate of one arrival rate while another was increasing. We will propose candidate theories in the discussion section of this paper.

The objective of our "lunchtime" change in both driver and rider agent demand and supply was to create drastic changes similar to that what would be expected in a major metropolitan area during this time period. What is critical to note is that this performance difference—this emergent pattern—is only seen under variable run conditions, and not stationary run conditions. Consequently, during a statistical analysis for our model's stationary run, we find that the mean, median and maximum financial gain (by all agents) during a 24 h period was not significantly different between driver agents employing a Voronoi movement versus random movement behavior. Wherein the variable ("Saturday") scenario run, the median, mean, and maximum were contrasted, with Voronoi movement outperforming random movement decisions on the

Table 10 A portion of the "Saturday" scenario that contains elements of the scenario portions where performance divergence was seen, yet, no divergence emerged for this sequence

10.30 PM	90	10	Low rider activity—riders are at their destinations. Drivers still on the road expecting a rush of new riders
11.30 PM	80	10	Some drivers give up, exit the simulation
12.30 AM	75	30	More drivers give up. Riders beginning to end their work shifts

aggregate; though the majority of the performance improvements came from the time-periods where Voronoi movement allowed a greater rate of customer pick-ups (1 PM and 3.30 PM). It was not immediately apparent that Voronoi movement outperformed random movement for every time-period of the variable run scenario.

4 Discussion

Clearly, the model shows that driver agent positioning prior to rider agent pickup influences the financial utility gained by driver agents (consequently this also means that riders receive more consistent pickups with shorter wait-times). But this emergent phenomenon is not recognizable unless a realistic variable arrival rate scenario is utilized. I argue that this is true because of the *spreading* nature of Voronoi movement. In highly volatile rider-supply and driver-demand areas of our variable scenario runs, spreading-out behavior ensures that drivers are more evenly distributed, and by being so they are better positioned to "catch" riders in a moment of higher demand. Where, if driver agents choose to continue moving and placing themselves randomly in moments of drastic change to supply and demand, their *catching* behavior is set to be limited and thus are unable to maximize their financial gain and adapt to their surroundings. This is the case for when both rates of entry for drivers and riders are increasing and decreasing simultaneously, thus it should be noted that opposite signs for the first derivative of the profit variable are not a requirement for this phenomenon to occur.

The details of when this occurred are also important. In the "Saturday" scenarios, Voronoi movement did not outperform random movement in all variance combinations. For example, Table 10 shows a portion of the variable scenario at the 10.30 PM time-period. Note the decrease in driver capacity throughout the listed time-periods from 90 to 75, while rider arrival rates remained constant and then increased to 30 per time unit. Figure 7 shows that in this time-period there was no divergence in performance between Voronoi and random movement, even though elements of both the rise and fall seen in the scenario portions (where visible drastic change was present) was also embedded in this particular sequence of agent arrivals.

Thus, we must then conclude that sudden increase and decrease in rider demand in conjunction with a steady or a slightly decreasing capacity (supply) provides a sudden spatial vacuum in the model's geographic configuration which is best enclosed by agents who are actively trying to move away from each other—a Voronoi movement pattern. But, that this pattern—that allows drivers to cover more space and 'catch' more riders—occurs only where there is enough drivers on the spatial geography such that there are actual additional riders who will be picked up by this movement. In other words, if decreasing or increasing variations in both agents' entry are occurring, Voronoi movement will provide drivers with an advantage over random movement, given that there are enough free drivers (without a passenger) and enough riders (without a driver) ready for pickup; an opportunity for maximizing utility must exist. After all, an increase in the spatial spread of driver agents allows for an increased probability of executing a pickup of a rider agent, but only when there are riders to be picked up.

The key to reproducing this pattern is that it must be part of a sudden and/or variable change scenario for arrival rates. We hypothesize that a stationary run will not emerge this phenomenon because, with unchanging arrival rates, incremental improvements will not allow for a critical mass of spatial imbalance in the location of riders and drivers.

There is a connected phenomenon observed in supply chain management theory that can be associated with this system property—what is known as the bullwhip effect. The phenomenon is widely understood as that of being a powerful reaction at the far end of a long supply chain which is often created from a small change in the point of origin of the chain. If the change is more sudden the effect is more compounded. In this case, the effect can be seen in the time-delayed spatial response of one agent group to another, not in a supply chain.

This behavior can be described as emergent. The pattern of maximizing utility through the prior, strategic positioning of driver agents is somewhat unexpected since all entry and exit of agents and their locations on the geography of the model are random. One might surmise (incorrectly, as we have shown) that if the random placement of rider agents and random placement of driver agents forms the core of the topological interactions of agents in this model, that through intuition alone there would be no clear gain in Voronoi movement behavior over random movement behavior. But as we have shown that while this is true for constant arrival rates, there is a difference in variable run conditions.

This emergent behavior does not seem to occur in the stationary runs of the model because at a constantly random and stable rate of entry for both agent types there is never a sudden vacuum to be capitalized upon.

This result provides researchers in this area with several important lessons: If one seeks to test movement behaviors of a ridesharing or an autonomous vehicle system, it should be tested under highly variable conditions in order to observe true emergent behavior. Stationary testing of spatial models would seem to be misleading and ineffective in this regard. Moreover, I propose that the testing of agent-agent interaction on any topological space for which there exists an entry or arrival dynamic would be subservient to the conclusions presented herein; though additional testing remains to conclude so irrefutably.

It's also paramount to realize that typical system dynamics modeling (differential equations) would most likely fail in producing the phenomenon as we have observed it, assuming a closed form equation for the mechanics of the model can be found, to begin with. Thus, we consider the experiment an added piece of important evidence for the utilization of agent-based techniques in autonomous vehicle and ridesharing service modeling and simulation (theoretically the difference between autonomous vehicle and ridesharing service modeling is negligible from the modeler's perspective).

5 Summary and Future Work

In this paper, we described the development of an agent-based model for ridesharing services in the Washington, D.C. area. The model simulates riders and drivers through simple interactions on an accurate data-driven geospatial configuration. This model forms the basis for a number of experiments and model extensions that could yield greater insights into the ridesharing economy as it develops, expands, and evolves.

Our conclusions showed the importance of running experiments utilizing agent-based modeling runs not only in the form of stationary runs but in the form of variable scenario runs designed to create unpredictable effects that can—and in our case did—yield greater insights which otherwise would not have been observed.

Specifically, we found the emergence of a pattern where prior positioning of driver agents had a significant effect on pickup rates, and thus on the financial gain (utility) of drivers. We also found that this pattern emerged from a simple spreading-out behavior, which we called Voronoi movement and that this movement pattern outperformed random movement patterns even with randomly distributed arrival rates for both agent types. However, this emergent phenomenon was not observable unless a variable scenario was utilized in the experimentation process.

Consequently, we showed that driver to driver agent interactions, which form a symmetrical information environment can provide increased utility for drivers, and consequently for the ridesharing service and riders as well in some cases.

Current operating procedures of the leading ridesharing services do not allow drivers to gain access to location information of other drivers, and thus do not allow for movement behaviors that are dependent on that additional information. The symmetrizing of information can yield greater utility for all sides of this equation, including service, rider, and driver. Perhaps ridesharing services believe that giving less information to drivers would allow them more centralized control which they can use to better optimize the ridesharing experience, but evidence that this is true is not without question if we consider the natural fluctuations in drivers' and customers' supply and demand. The question posed by us here is whether more information may allow drivers to create adaptive and cooperative strategies to maximize their financial gain and by consequence, all other parties. We showed that this may be the case for one simple behavior and we demonstrated the conditions under which future spatial behavioral testing should be implemented if we are to be confident in the outputs of our simulations.

There are many pathways that this model can take—going forward. Primarily, the most interesting extension would be to add more spatially complex behaviors in agent-to-agent interactions and then to observe the results. It is not trivial that we test expected utility for an information symmetry scenario with only one movement-type behavior. More spatial movement patterns grounded in expected behaviors should be tested to quantify the difference in a service that allows more drivers to have more information and one that does not. There are improvements to be made in the spatial configuration of the model itself as well. For example, the inclusion of spatial demographics to enrich probability distribution calculations, adding road direction and traffic patterns, as well as utilizing more detailed geographic datasets would all make significant improvements to the model's efficacy and predictive power.

References

1. Anderson, D.N.: Not just a taxi: for-profit ridesharing, driver strategies, and VMT. Transportation **41**(5), 1099–1117 (2014). https://doi.org/10.1007/s11116-014-9531-8
2. Bash, E.: Dynamic ridesharing: an exploration of the potential for reduction in vehicle miles traveled. Ph.D. Proposal (2015)
3. Cho, S., Yasar, A.U.H., Knapen, L., Bellemans, T., Janssens, D., Wets, G.: A conceptual design of an agent-based interaction model for the carpooling application. Procedia Comput. Sci. **10**, 801–807 (2012). https://doi.org/10.1016/j.procs.2012.06.013
4. Cho, S., Yasar, A.-U.-H., Knapen, L., Patil, B., Bellemans, T., Janssens, D., Wets, G.: Social networks in agent-based models for carpooling. In: Transportation Research Board 92nd Annual Meeting, pp. 13–2055 (2013)
5. Furuhata, M., Dessouky, M., F., Brunet, M. E., Wang, X., Koenig, S.: Ridesharing: the state-of-the-art and future directions. Transp. Res. Part B: Methodol. **57**, 28–46 (2013). https://doi.org/10.1016/j.trb.2013.08.012
6. Hussain, I., Knapen, L., Galland, S., Yasar, A.-U.-H., Bellemans, T., Janssens, D., Wets, G.: Agent-based simulation model for long-term carpooling: effect of activity planning constraints. Procedia Comput. Sci. **52**(Ant), 412–419 (2015). https://doi.org/10.1016/j.procs.2015.05.006
7. Knapen, L., Keren, D., Yasar, A.U.H., Cho, S., Bellemans, T., Janssens, D., Wets, G.: Analysis of the co-routing problem in agent-based carpooling simulation. Procedia Comput. Sci. **10**(270833), 821–826 (2012). https://doi.org/10.1016/j.procs.2012.06.106
8. Knapen, L., Keren, D., Yasar, A.-U.-H., Cho, S., Bellemans, T., Janssens, D., Wets, G.: Estimating scalability issues while finding an optimal assignment for carpooling. Procedia Comput. Sci. **19**(Ant), 372–379 (2013). https://doi.org/10.1016/j.procs.2013.06.051
9. Martin, C.J.: The sharing economy: a pathway to sustainability or a nightmarish form of neoliberal capitalism? Ecol. Econ. **121**, 149–159 (2016). https://doi.org/10.1016/j.ecolecon.2015.11.027
10. Nielsen, J.R., Hovmøller, H., Blyth, P.L., Sovacool, B.K.: Of "white crows" and "cash savers:" a qualitative study of travel behavior and perceptions of ridesharing in Denmark. Transp. Res. Part A: Policy Pract. **78**, 113–123 (2015). https://doi.org/10.1016/j.tra.2015.04.033
11. Nourinejad, M., Roorda, M.J.: Agent-based model for dynamic ridesharing. Transp. Res. Part C: Emerg. Technol. **1**, 117–132 (2015). https://doi.org/10.1016/j.trc.2015.07.016
12. Pelzer, D., Xiao, J., Zehe, D., Lees, M.H., Knoll, A.C., Aydt, H.: A partition-based match making algorithm for dynamic ridesharing. IEEE Trans. Intell. Transp. Syst. **16**(5), 2587–2598 (2015). https://doi.org/10.1109/TITS.2015.2413453

13. Rayle, L., Shaheen, S., Chan, N., Dai, D., Cervero, R.: App-based, on-demand ride services: comparing taxi and ridesourcing trips and user characteristics in San Francisco. Univ. Calif. Transp. Cent. **94720**(August), 1–20 (2014). https://doi.org/10.1007/s13398-014-0173-7.2
14. Rayle, L., Dai, D., Chan, N., Cervero, R., Shaheen, S.: Just a better taxi? a survey-based comparison of taxis, transit, and ridesourcing services in San Francisco. Transp. Policy **45**, 168–178 (2016). https://doi.org/10.1016/j.tranpol.2015.10.004
15. Shaheen, S. A.: Shared mobility innovations and the sharing economy. Transp. Policy, 1–2 (2016). https://doi.org/10.1016/j.tranpol.2016.01.008
16. Shaheen, S. A., Chan, N. D., Gaynor, T.: Casual carpooling in the San Francisco Bay Area: understanding user characteristics, behaviors, and motivations. Transp. Policy, 1–9 (2016). https://doi.org/10.1016/j.tranpol.2016.01.003
17. Tran, Q. T., Tainar, D., Safar, M.: Transactions on Large-Scale Data- and Knowledge-Centered Systems. p. 357 (2009). ISBN 9783642037214
18. Voronoi, G.: Nouvelles applications des paramètres continus à la théorie des formes quadratiques. Premier mémoire. Sur quelques propriétés des formes quadratiques positives parfaites. Journal Für Die Reine Und Angewandte Mathematik (Crelle's Journal), 1908 **133** (1908). https://doi.org/10.1515/crll.1908.133.97
19. Wilensky, U.: NetLogo. Center for Connected Learning and Computer-Based Modeling, Northwestern University, Evanston, IL (1999). http://ccl.northwestern.edu/netlogo/
20. Zargayouna, M., Zeddini, B., Scemama, G., Othman, A.: Agent-based simulator for travelers multimodal mobility. Front. Artif. Intell. Appl. **252**, 81–90 (2013). https://doi.org/10.3233/978-1-61499-254-7-81
21. Zolnik, E.J.: The effect of gasoline prices on ridesharing. J. Transp. Geogr. **47**, 47–58 (2015). https://doi.org/10.1016/j.jtrangeo.2015.07.009

Stigmergy for Biological Spatial Modeling

Megan Olsen

Abstract Complex adaptive systems can be characterized as systems that are comprised of groups of agents following simple rules that, collectively, produce emergent, complex behavior. The key to these emergent properties are the interactions—the exchanges of information—between the agents. Many biological systems can be studied using a complex adaptive systems approach, such as colonies of bees or ants. In some of these biological systems, the communication between individual agents is indirect. This type of communication is termed stimergy: a relatively small amount of information being shared through the environment, rather than directly from agent to agent. This information is nonetheless crucial to the self-organizing properties of the system, and is used by the agents to inform decision making, such as when ants follow a trail of pheromones left by other ants. In this chapter we describe computer simulations of two such systems, created and used to conduct experiments on various types of stimergy: collaboration within a predator-prey system, and angiogenesis in cancer growth. The first utilizes a cellular automata model, and the second a multiscale agent-based model. Further, this paper defines various options of communications for these simulations, and examines the use of similar communication paradigms in these two different types of models. Results support that stigmergy can be adapted to a variety of situations. Also, that awareness of the speed of algorithmic decisions can increase its usefulness in biological modeling. These ideas can be adapted to many other modeling situations other than the classic examples of self-organization like bees or ants.

1 Introduction

The term *Complex Systems* has been used in a variety of ways by scientists, often referring merely to a system that is not simple. However, the field of Complex Systems studies a much less general class of problems that are complex in ways that may include simplicity at its heart: emergent, adaptive, nonlinear, chaotic, and multi-

M. Olsen (✉)
Loyola University Maryland, Baltimore, MD, USA
e-mail: mmolsen@loyola.edu

T. Carmichael et al. (eds.), *Complex Adaptive Systems*, Understanding Complex Systems, https://doi.org/10.1007/978-3-030-20309-2_8

scale complexity, to name a few [8, 51, 53]. This field is inherently interdisciplinary, studying much of the world around us using techniques from a variety of disciplines.

Emergent behavior refers to the whole as being more than just the sum of the parts, for example, when simple rules create complex dynamics. Anderson summarizes this idea by stating that "the reductionist hypothesis does not by any means imply a 'constructionist' one" [4]; i.e., we cannot necessarily construct a complex system purely by understanding its parts. However, learning what creates this complexity within systems, as well as how these complex systems work and interact with each other, can further our understanding of many fields and of science in general. We want to be able to anticipate the behavior that emerges from these small pieces being combined, even if that behavior does not seem fully described by those pieces.

Self-organization is often a noticed outcome of emergent behavior, and a prime example of complex behavior appearing from simple rules of interaction. It can be seen in many biological systems, including molecular or cellular biology, dynamics of species in ecology, or human behavior. A popular example is swarm intelligence, which refers to the natural organization of bees, ants, or flocks of birds without centralized communication [8, 12, 55]. In this case, self organization through simple interactions are causing a large coordinated effort for travel, finding food, or general communication [56]. This phenomenon is also observed in computer networks. Probably the best-known example is the Internet, a network that is self-organized but yet still follows a power law in link organization [44]. Arguably the most commonly known self-organization in a computing system is Conway's Game of Life, in which simple rules within a cellular automata create interesting patterns [21].

Self-organization is generally accomplished via communication among agents or entities within a system. Often our first inclination is to consider direct communication between agents, but this approach is not the only option. Communication can instead be indirect, where only a relatively small amount of information is shared through the environment and then used to inform entity decisions. This approach is called *Stigmergy*. Stigmergy is used by ants and bees as a way to enable self-organization. It can also be used for human collaboration (such as Wikipedia), or to encourage collaboration and cooperation in computing systems. It has been shown that only a small amount of information from a few informed individuals is necessary for effective group decision making [14]. One of the benefits of stigmergy is that it is not necessary for the entities communicating to be aware of each other, or know much about each other. This type of communication can be used effectively in many spatial models for both representing activities known to use stigmergy in the real world, and for representing systems that at an abstract level act in this way even though the actual mechanisms are different. Figure 1 displays three basic approaches to communication within an individual-based model. Direct communication with another entity in the system (Fig. 1c), either restricted by distance or not, is not a form of stigmergy. Diffusion of information (Fig. 1a) is incredibly common in self-organization, and can vary based on its parameters such as how fast diffusion occurs, how fast the information decays, and the amount of information diffused. Individuals can also leave a trail of information such as how ants leave pheromone trails (Fig. 1b). These approaches can also be combined and adapted to fit the given problem, such as a trail

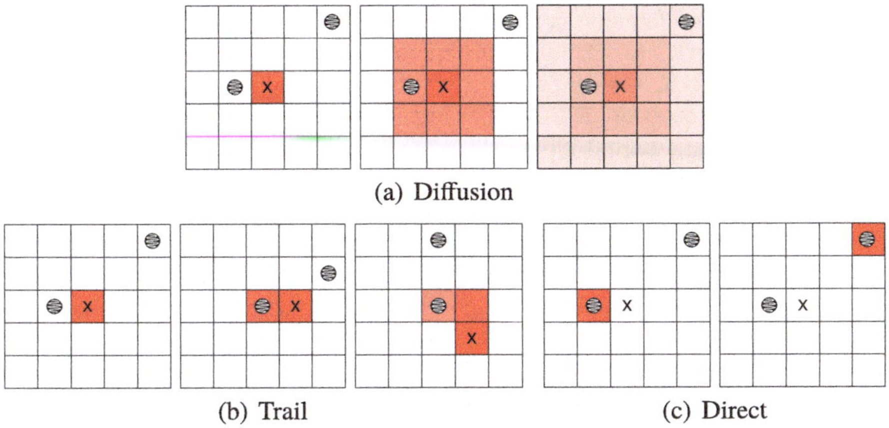

Fig. 1 Examples of communication within an individual-based model: **a** Information is diffused into the the environment, **b** A trail is left by the individual, and **c** another individual is directly contacted. The X represents the communicating individual, and the circles represent other individuals who may obtain the communicated information. A darker color represents a stronger message. Only **a** and **b** are forms of stigmergy

that also diffuses, or a bee's waggle dance that it essentially a limited diffusion of information that ceases once the dance ends. As will be examined in this chapter, our choice of communication details can affect the results we obtain from running our model, and thus we must be thoughtful in our choice of communication paradigm when designing a system.

Systems that exhibit emergent behavior and self-organization may also interact on *multiple scales*, as do systems that we do not traditionally call emergent. Many systems in both biology and computer science work on multiple scales, including the scales of time and space. Cellular biology is a prime example of a multi-scale system: even if we only look at what controls the cell, and ignore the overall interactions at a cellular level, there are different time scales for replication and interaction, and different sizes for proteins, RNA, etc. Thus, studying cellular biology can be done by examining cells at the cellular level, the subcellular level, or the tissue level, for example. At the same time, it may be beneficial to analyze the overall functioning of the plant or animal created by these cells, or even the interaction of animals within a species or between species. Thus, every organism is representing multiple space/time scales of interest, and to truly understand the mechanisms at work it is necessary to analyze multiple scales at once. This leads to multi-scale models, both mathematical and computational.

Modeling of biological systems allows us to analyze and approach them in a different way than is traditionally done by biologists. Ideally, biologists and modelers work together to form a complete cycle where biological experiments feed into computational/mathematical models, which then feed back into the biological experiments to continue the cycle [28]. Both fields are still working to achieve this collaboration and balance. However, many modelers are either also involved in the traditional "wet" biological experiments or work with biologists and their data to ensure that the created models are biologically validated, and can therefore have the potential to feed back into the biological experimental cycle. Ideally, models are used for examining many scenarios, as they allow the study of many scales of the system together in a way that is often difficult or impossible in a traditional lab setting.

There are many different types of models used to study biological problems: ordinary differential equations, partial differential equations, agent-based simulations, event-based simulations, cellular automata, computational geometry, and more. Non-linear dynamic approaches examining self-organizing systems are also used for multi-scale problems such as cancer [11]. The key is to determine the appropriate approach for the subfield and specific problem being studied. In this paper we focus on spatial models, specifically cellular automata and agent-based models. These models are also known as individual-based models. In each of these models there are individuals making decisions within the model, that affect the overall outcome. In cellular automata, each location or cell within the multi-dimensional world follows a set of rules each time step. In agent-based models, individual agents follow rules governing their behavior, and may move within the world. Cellular automata have been used for modeling for many decades, although agent-based models have become the more popular approach in recent years.

To show how stigmergy can be utilized in a variety of modeling situations, in this chapter we'll discuss modeling two different complex systems using stigmergy, with two different modeling paradigms: a cellular automata model of predator-prey interaction with computational emotions [41], and an agent-based model of cancer cell growth [38]. For predator-prey, we create a cellular automata model to study the use of emotions in prey collaboration. To study cancer cell growth we create an agent-based model of tumors with angiogenesis, to enhance our understanding of the role angiogenesis plays in tumor growth. Although they seem initially different, both utilize stigmergy for self-organization, although they utilize stigmergy in different ways. Cancer is a multi-scale model on a three-dimensional grid, whereas predator-prey is on a two-dimensional grid. Similar principles can be used to study each problem, and we'll discuss the variations on stigmergy used to model each one. We will focus on two-dimensional and three-dimensional grid worlds in this chapter, but the ideas presented here can be expanded to different spatial systems. In the next section we'll discuss the predator prey model, followed by a section on the cancer model, and ending with a discussion section.

2 Cellular Automata Model of Predator-Prey

Collective behavior can refer to both human and animal tendencies to influence each other's behavior. In this section we are most interested in collective animal behavior that leads to improved species survival within a predator-prey system. A major topic of population dynamics, the study of the development of either a single or multiple interacting species, is the cycling of predator and prey populations. Predator-prey dynamics relate to a wide variety of ecological situations, from microbial phagocytosis to lions and gazelles. Most often predator-prey systems are built to describe animal species, with at least one species as prey and one as predator; however, they are not limited to describing only two species. The Lotka-Volterra equations are based on the classic logistic equation, and commonly used to model this type of interaction [31]. However, it has been argued that these equations are not sufficient for modeling natural phenomena [29].

Cellular Automata (CA) offer a popular mechanism to analyze population dynamics as they directly represent spatial interactions between entities [26]. CA allow the creation of rules for determining how an entity will interact with its neighbors. CA can give rise to complex dynamics via simple rules, which enhances its desirability for modeling complex phenomena such as collective behavior, assuming that the appropriate simple rules can be designed or learned. In population dynamics CA models, entities explicitly exist on a grid and interact with neighbors. In predator-prey models, multiple interacting species move around the grid, with predator chasing prey. In some models, movement is purely defined by the birth of new entities into neighboring cells [15, 19], whereas other models allow individuals to actively move around the grid [16, 25]. We utilize both approaches.

For predator-prey dynamics in a CA it is useful to analyze the patterns created in the system, including how the number of predators and prey fluctuate in the system over time, as well as how mixed they are spatially [15, 25]. Results from [15] indicate that a combination of Lotka-Volterra [31], an individual's ability to change, and the spatial structure of the CA give rise to both predators and prey self-organizing into self-sustaining patterns. It has also been proposed that taking the environment into account significantly affects results, potentially making them more realistic as living creatures naturally have outside influences other than a predator or prey [19].

We propose that emotions can be used to increase collective behavior in a population dynamics model. Although there are a number of human psychological theories of emotion [18, 49, 50], it is generally agreed that emotions serve the purpose of increasing our ability to interact with our environment in a successful manner. Rolls has argued that human emotions have ten functions, including reflexive behaviors and motivation [50]. It has been suggested that human emotions were initially evolved due to the need to survive, showing many commonalities with reactions of prey and predator [32, 49]. Although there have been arguments that any defensive action in a prey represents emotions, more recently that view has been modulated to instead argue that although this may not always be the case, it is still likely that initial reactions to threat in animals can precede emotions [7].

Our predator-prey model is an analysis of predators that feed on prey that feed on a stationary food source. For collective behavior to arise, information in the form of emotions is shared between conspecifics (individual of the same species), based on the idea that communication of emotions between humans is an important aspect of our interactions. Both prey and predator make decisions with their personal emotions and the shared emotions taken into account. Emotions are affected by environmental events, and thus represent a high level of information about the environment. Although emotions may initially seem like an odd form of information to share, the development of emotions in higher animals has been conjectured to originate for purposes of survival in basic scenarios such as predator-prey [7, 32], and thus emotionally-inspired rules are a natural extension.

We choose to include the six basic universal emotions as defined by Ekman for our prey and predator: happiness, sadness, fear, anger, disgust, and surprise [18]. Emotions occur in response to specific world events, such as the happiness of food consumption and the fear of predator encroachment. The emotional state of a member of a species can be communicated to a member of the same species and thus affect their emotional state, with the goal of increased coordination and cooperation.

We consider this approach to emotional communication as an efficient way of transferring information that is crucial for the survival of the group, and analyze two different techniques for sharing emotion: local sharing from neighbors, and sharing by a decaying trail similar to ant pheromones [37]. Collective animal behavior is often studied from a few different perspectives: examining biological causes of collective behavior in animals [37], examining how to build models to capture collective behavior [13], and being inspired by collective behavior [9]. Often these perspectives overlap, as more recent interdisciplinary efforts use modeling to help decipher the biological phenomena and create a better understanding of what underlying mechanisms can result in various forms of collective behavior. Our goal is to determine which modeling approach leads to the best collaboration within a species. Groups of prey in real situations will exchange information about their surroundings, increasing their likelihood of survival [54]. Although our individuals are inherently selfish in that they make the best decision for themselves, our results show that communication of emotion can increase collective behavior for both predator and prey, i.e. self-organization through stigmergy. This work is based on our previously published studies [23, 24, 39].

2.1 Our Model

We use a four-species (prey, predator, stationary food which is considered the prey's prey, empty space) two-dimensional cellular automata (CA) model to examine predator-prey dynamics in an environment that includes disease, and where emotions are developed individually and communicated to neighboring conspecifics. Each point in the model may hold a single entity from any species at any given point in time. Each of the non-empty species are able to reproduce and die, and predator and prey can also move. Our grid structure is defined as a torus, and thus there are no corners requiring special treatment. Each individual's next step is determined based on probabilities and their neighbors within the Moore neighborhood.

A predator eats a prey when it moves to the prey's position. Once all of a species have been eaten, no more of that species can come into being in the system. However, as long as there is still at least one of a species alive it is possible for it to reproduce to create a new one, as for simplicity reproduction is asexual. Although this is a simplification of reproduction, it will not seriously impact the study, and is a commonly used abstraction.

Predator and prey can exist either with or without emotions. Food is not affected by emotions. We will first describe the model without emotions, and then describe how emotions are generated and communicated through stigmergy, and then how they modify individual behaviors.

2.1.1 Probabilistic and Neighbor-Based Rules

At each time step, each cell occupied by food will update following these rules:

- Reproduction: If it is at maturity age it produces a new food item into a vacant adjacent square, if one exists. Age is reset to zero.
- Disease: When a new food item is produced it has a 0.1 probability of being diseased. If the parent is diseased, the probability doubles. Disease lasts for a minimum of 2 timesteps. After that minimum time has elapsed, there is a fixed probability *Cure* of disease being cured each timestep.

Predator and prey follow slightly more advanced rules at each time step:

- Reproduction: If it is of maturity age there is a fixed probability *Rep* of a new predator/prey being created in a vacant adjacent square, if one exists. Age resets to zero after reproduction.
- Disease: If a predator eats a diseased prey, the predator becomes diseased. Disease lasts a minimum of 2 timesteps, after which there is a probability *Cure* of disease being cured each timestep.
- Movement: Movement occurs by computing a local gradient of preference for the surrounding cells (Eq. 1) that biases their otherwise random movement. Individuals can move either to an empty cell or a cell with their food source.

$$g_{d,sp}(t,x,y) = food_{sp}(t,x_d,y_d) \tag{1}$$

  where d in $D = \{NW, N, NE, W, E, SW, S, SE\}$, t is the current time, $g_{d,sp}(t,x,y)$ is the preference for a d at time t for an individual at location (x,y), $sp \in \{\text{predator,prey}\}$, and $food_{sp}(t,x_d,y_d)$ returns the existence of prey for a given species sp at locations in direction d.
- Movement rate: When diseased, movement probability decreases (*DiseaseMove*).
- Hunger: Hunger increases by 1 each timestep it doesn't eat, or by 2 if it is diseased. Hunger is decreased to zero when it eats.

2.1.2 Individual Emotions

Emotions are calculated at the end of the sequence described in the previous subsection, and are used by predator and prey when determining the next movement direction as well as their probability of reproducing. They are based on Ekman's original six basic emotions (fear, anger, sadness, happiness, disgust, surprise) [18].

Table 1 Calculation for emotion based on experiences

Emotion	$X_{e,sp}(t, x, y)$
Happiness	1 if ate prey, 0 otherwise
Sadness	t [timestep of last reproduction]
Anger	exp^{hunger}
Fear (predator)	Anger of neighboring predators
Fear (prey)	Number of neighboring cells with predators
Disgust	1 if ate diseased prey, 0 otherwise
Surprise	$\sum_e \frac{E_e(t,x,y)-E_e(t-1,x,y)}{5}$, where e does not include surprise

Each individual maintains their own emotions. Emotions are independent for each species, and each emotion is affected by different experiences related to neighbors, hunger, reproduction, and disease. The numerical values used to represent these experiences when calculating emotions can be seen in Table 1. Individuals also communicate their emotion, which can influence the emotions of other nearby individuals from their own species. Communicated emotions of one species cannot be seen or interpreted by the other species.

An individual's emotions are based on experiences, as well as their previous emotion and the emotions being communicated nearby. They are computed at each time step as seen in Eq. 2. The previous emotion and communicated emotion are both discounted, to prevent them from overpowering newer experiences or causing monotonically increasing emotions over time.

$$\begin{aligned} E_{e,sp}(t+1, x, y) = (1 - c_{m,sp})(X_{c,sp}(t, x, y) + c_{c,sp} * CE_{e,sp}(t, x, y)) \\ + c_{m,sp} * E_{e,sp}(t, x, y) \end{aligned} \tag{2}$$

where $e \in \{\text{fear, anger, sadness, happiness, disgust, surprise}\}$, $X_{e,sp}(t, x, y)$ is the unique experience of each emotion for each species (Table 1), $c_{m,sp}$ is the memory discounting coefficient that determines what percent of the new emotion is based on new versus old emotional information (Eq. 3, bounded to [0.1,0.5]), $c_{c,sp}$ is the discounting coefficient for communicated emotion (bounded to [0,1)), and $CE_{e,sp}(t, x, y)$ is the communicated emotion at position (x, y) at time t (Eqs. 4 and 5).

$$c_{m,sp} = 0.1 + 0.4 * \frac{abs(E_{surprise,sp}(t, x, y) - E_{surprise,sp}(t-1, x, y))}{E_{surprise,sp}(t, x, y) - E_{surprise,sp}(t-1, x, y)} \tag{3}$$

2.1.3 Emotion Communication

Both predator and prey will send their emotions to the environment, similar to ant pheromones. These markers are only understandable by entities of the same species. After each timestep, all emotions are decayed (both internal and environmental). We hypothesize that this stigmergy will help each species' survival through cooperation.

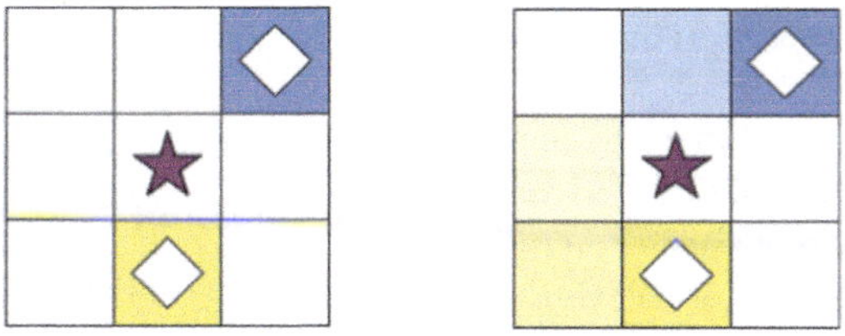

Fig. 2 The two communication styles. The star represents where an entity currently exists who is receiving communicated emotion through the environment. The colored background represents the currently shared emotion that is used by the entity to modify their emotions, and the diamonds represent the entity whose emotion the star is receiving. **a** Local communication in which only the current neighbor emotional state influences the entity. **b** Trail communication in which a decaying trail is left by the diamond

Local Stigmergy

In this case, only emotions from the immediate neighborhood from the previous time step are used in the decision (Fig. 2a). Additionally, emotions are only read from the map if a conspecific existed at that point; i.e., only emotion shared from events in the previous timestep are included. When an individual is determining their own emotions for timestep (t+1), they will calculate what emotions are being shared through the environment via Eq. 4 from timestep t.

$$CE_{e,sp}(t,x,y) = \sum_{d \in D} S_{sp}(t, x_d, y_d) * E_{e,sp}(t, x_d, y_d) \tag{4}$$

where $S_{sp}(t, xd, yd)$ is 0 or 1 denoting the existence of species *sp* and $E_{e,sp}(t, x_d, y_d)$ is the amount of emotion in direction *d* from position (x, y) at time t.

Trail Stigmergy

With stigmergic communication, the shared emotions will remain at the cell over time and will decay with each timestep (Fig. 2b). When an entity moves to that cell and leaves its own emotional mark, the newly shared emotions will be combined with the previously shared emotion, affecting calculation of an individual's new emotional state (Eq. 5). Although the individual emotion always decays linearly, we test two different forms of decay for the communicated trail emotion: linear and geometric. This variety of decay examines whether the rate of decay affects either species' ability to self-organize.

$$CE_{e,sp}(t,x,y) = \sum_{d \in D} E_{e,sp}(t, x_d, y_d) \tag{5}$$

2.1.4 Rules Enhanced by Emotions

Predator and prey with emotions have altered reproduction rates and movement direction preferences. Other aspects are calculated the same as previously defined. As emotions are used to encode and communicate various features of the environment to modulate the behavior of individuals, the components of the environment incorporated into emotions are intuitively useful for survival, and should thus be beneficial for each species.

Reproduction rates are altered positively by happiness, and negatively by disgust and anger. Additionally, prey will not reproduce while their fear is above a threshold. The calculation of reproduction rate $R(t)$ can be seen in Eq. 6,

$$\begin{aligned} R(t) = {} & Rep * (1 - Rep) * (E_{happy}(t, x, y) \\ & - Rat * E_{disgust}(t, x, y) - (1 - Rat) * E_{anger}(t, x, y)) \end{aligned} \tag{6}$$

where Rep is the initial probability of reproducing after the maturity age has been reached, Rat is the ratio of how much disgust versus anger decreases reproduction, and $E_{happy}(t, x, y)$, $E_{disgust}(t, x, y)$, and $E_{anger}(t, x, y)$ represent current emotional values.

The local preference gradient for movement considers emotions, where an individual will move toward the highest positive value $g_{d,sp}(t, x, y)$. This differential is taken to be the difference between the emotion in a given direction and the current emotion of the individual (Eq. 7),

$$\begin{aligned} g_{d,sp}(t, x, y) = {} & food_{sp}(t, x_d, y_d) \\ & + \sum_e valence_e * (E_{e,sp}(t, x_d, y_d) - E_{e,sp}(t, x, y)) \end{aligned} \tag{7}$$

where (x, y) represents the individual's current location, (x_d, y_d) represents locations in direction d, e is taken over all emotions except surprise, and $valence_e$ is -1 for a negative emotion (fear, anger, sadness, disgust) and 1 otherwise.

2.2 Results

The simulation is run on a grid world of size 50 by 50. Simulations are run for a total of 2000 time steps with an initial random placement of individuals on the grid. Initial population sizes for predator, prey, and food were 240, 500, and 1500. Initial parameter searches were done on both population parameters and emotion parameters for all populations to determine which parameters to investigate further (Table 2). Each of the twenty initial placements are tested on the four emotion scenarios: no emotions, only one species using emotion, and both species using emotions. For the trail communication we test two types of decay, each with two emotion decay

Table 2 Simulation parameters

	Food	Prey	Predator
Rep	n/a	0.75	0.75
Maturity age	2	6	7
Cure	0.8	0.8	0.8
Diseasemove	n/a	0.8	0.8
Starvation	n/a	2	6
Rat	n/a	0.4	0.4
Fear threshold	n/a	n/a	0.5
$c_{c,sp}$	n/a	0.5	0.5

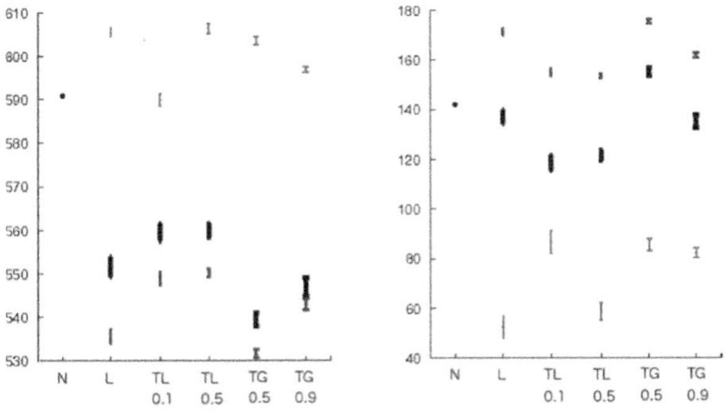

(a) Prey Population (b) Predator Population

Fig. 3 Average prey (**a**) and predator (**b**) populations for each parameter combination. The x-axis shows the decay values for each type of communication (N: no emotion; L: local; T: trail (L=linear, G=geometric)). Black (darkest) represents when both species use emotion, red (second darkest) represents when only predators use emotion, and blue (lightest) represents when only prey use emotion

rates: linear (0.1,0.5), and geometric (0.5,0.9). The graphs show the averages over all experiments, with standard deviation denoted.

We expect that high decay rates in trail communication will lead to population averages that are almost indistinguishable from the local communication case, as emotions would not be able to linger long enough to be significantly different from only counting the previous time step. There is not enough similarity between high decay versus low decay to say this is the true effect on populations (Fig. 3). Most likely there are other changes occurring that affect experiences and decisions due to the emotions; perhaps just a small amount of lingering emotion can make enough of a difference to change results.

Population averages are consistently better for each species when they are the only species using emotion, but are affected by communication approach (Fig. 3). Predators benefit more from geometric decay than linear decay, whereas prey benefit the most in all cases except low linear decay. The rate of linear decay is irrelevant for predators, but only when both species use emotion. A potential cause of this result can be seen in the surprise and fear of prey (Fig. 4). The increase in surprise at a high geometric decay will cause prey to focus more on the present than the past, while a higher level of fear will cause prey to flee predators.

The change in populations indicates that prey are at a disadvantage if predators use emotion, whether or not they themselves do. However, predators are at an advantage if they use emotion even if prey also do, as long as the communicated emotion is decayed geometrically. Thus, the length of time a shared marker lasts in the environment may determine whether or not stigmergy is worth evolving.

We examine the actual emotions of the prey and predators to better understand the population fluctuations. If experiences are not changing due to emotion then we expect to see that a slower decay rate of emotion leads to higher overall emotional values for both species on average as they are only being overwhelmed by communicated emotion. However, for most of the emotions we do not see this type of trend (Fig. 4). Therefore, each communication approach is modifying behavior such that changes other than decay rate are causing the emotion levels, thus implying changing experiences and behaviors.

For both species a higher geometric decay rate with trail communication leads to more emotion than a lower geometric decay rate; this difference must lead to increased emotion from experience, since it is not due to increased communication. A linear decay affects each species differently, however. For prey, trail communication with a lower linear decay always leads to higher emotion than with a higher linear decay, as we would expect if communicated emotions are dominating their emotion calculations. For predators, both linear decay rates lead to the same level of anger, disgust, and happiness but an increased linear decay rate leads to decreased sadness and increased fear for predators. Therefore, with linear decay predators' individual experiences are more affected than prey experiences.

Local communication leads to the highest levels of prey anger and happiness. It also leads to lower predator anger and fear but higher sadness, disgust, and happiness than a linearly decayed trail communication. Since anger and happiness are both related to hunger in inverse ways, it is not surprising that the predator emotion trends between the two emotions are essentially opposing each other.

All prey emotions except anger and sadness are strongest when both species use emotion across all communication approaches (Fig. 4a–f). The increase of fear in prey when predators also use emotion is of particular interest, as it supports the idea that predator emotion improves their performance. Prey anger is increased when only prey use emotion, but prey sadness doesn't change; emotions therefore either do not seem to help prey find food, or are more effective in having them evade predators which may mean less time for eating. All independent predator emotions except fear are the same when either only predators use emotion or both species use emotion. It therefore seems that predators' use of emotions is not hindered or helped by prey modifications.

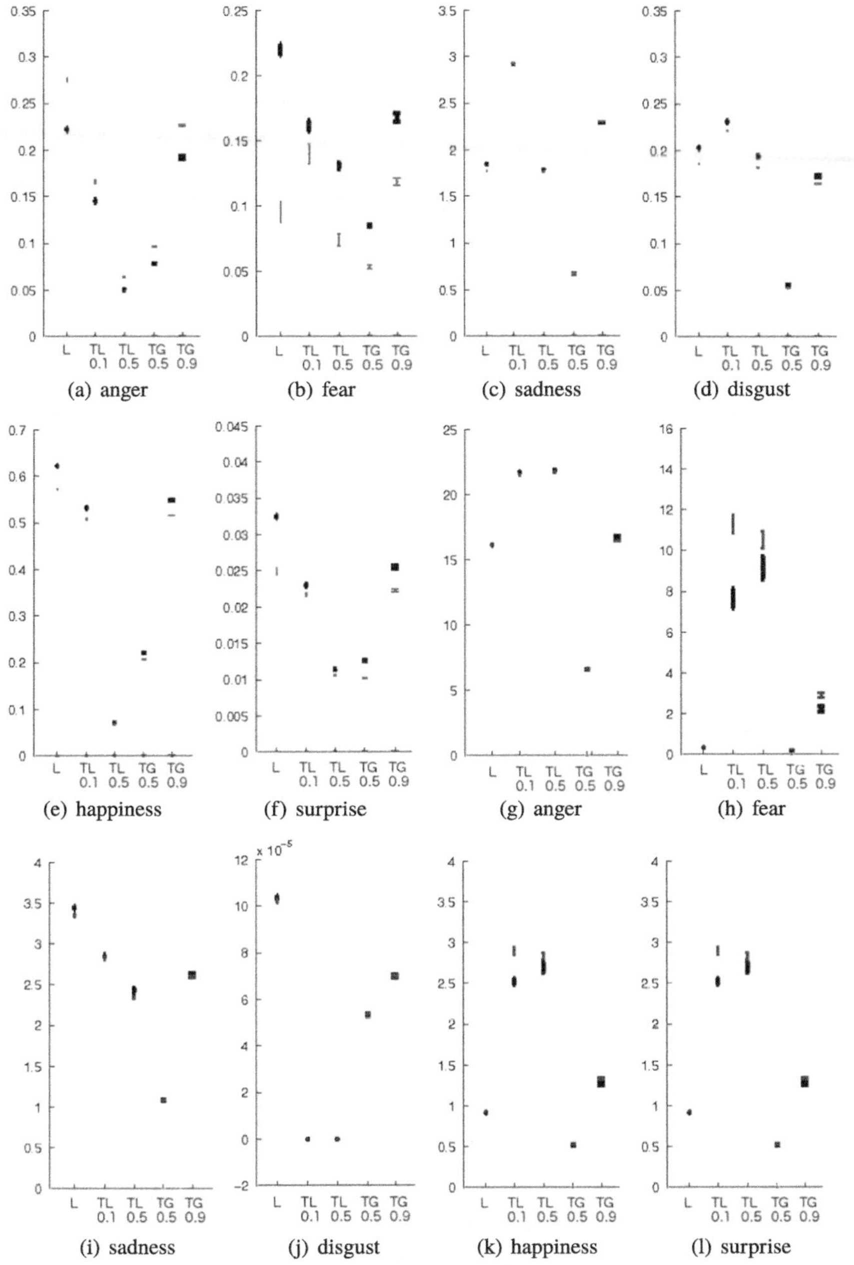

Fig. 4 Average individual prey (**a–f**) and predator (**g–l**) emotion for each parameter combination. The x-axis shows the decay values for each type of communication (L: local; T: trail (L = linear, G = geometric)). Black (darkest) represents when both species use emotion, red (second darkest) represents when only predators use emotion, and blue (lightest) represents when only prey use emotion

2.3 Conclusions

We have analyzed the use of computational communicated emotions toward increasing collaboration and collective behavior for both predators and prey in a Cellular Automata predator-prey model. Both species were given emotions inspired by Ekman's six basic universal emotions and current research on how emotions affect animals in predator-prey scenarios. Two methods of stigmergy were tested: local communication, and trail communication.

From the results we see that both species are benefited the most if they are the only species using and communicating emotion. However, we also see that the predator population benefits when both populations use emotion if using trail communication with a geometric decay. Thus, the predator is able to act collectively with a trail of information when the trail remains for a longer period of time. This result is logical, as a predator being able to detect prey from a further distance will likely be a better hunter.

Overall, communicated emotion can play a role in collective behavior for both predators and prey. Interestingly, the communication paradigm best for one species may be the worst for the other species. It will be interesting in the future to develop a system in which each species can evolve its communication strategies to see how the population dynamics and collective behavior are both affected by this additional dynamic. This type of system has already begun to be examined via reinforcement learning in predator prey [42].

3 Agent-Based Model of Tumor Angiogenesis

Cancer incidence is expected to rise worldwide from 12 million new people affected annually in the year 2000 to an anticipated 20 million in the year 2030, highlighting the urgent need to identify highly effective preventative and therapeutic interventions. More people are being diagnosed each year (even when adjusted to a standard population size). The only disease that is currently more deadly than cancer is heart disease. The overall goal of cancer research is to either eradicate the disease or make it no longer deadly. This goal is broken into many smaller high-level goals such as understanding how cancer cells are formed, how a single cancer cell develops into a tumor, what regulates the growth of cancer cells, how different types of cancer differ, how to better test for cancer in a patient, and how to remove cancer once it is discovered. The fields of Computational Biology and Systems Biology work toward a better understanding of the mechanisms behind this disease in a way that standard biological techniques cannot yet achieve. With the use of computational and mathematical models we can examine how biological systems work in combination, as well as create new testable hypotheses [58]. These models must be based on biological knowledge and data and validated before their results can be used to direct biological research. Although there is great promise from computational and mathematical techniques for cancer research, the cancer biology community is still in the process of learning to accept the potential of these models to contribute to the field [57].

To understand how to model cancer it is first necessary to have a better understanding of the disease itself. Cancer is essentially a disease in which certain cells no longer follow the originally defined genetic rules. There are a number of aspects to biological cells that are important for sustaining a cellular system such as a tissue. Cells are not only controlling their own behavior, but through inter-cellular communication they are also controlling the behaviors of their neighbors. Additionally, there are many larger aspects within the body that affect the cells, such as nutrient diffusion. Hanahan and Weinberg defined a set of six properties that represent the failures needed within a cell for it to become cancerous [22]. All of these modifications to the cell and tissue are necessary for a metastatic tumor cluster to form [22].

However, the order in which these failures occur is also important for a cell to become cancerous. These failures occur via mutations to the cell's DNA, which can occur due to a variety of causes including during proliferation, i.e. the creation of a daughter cell. Thus, losing both the ability to repair genetic mutations and the ability to undergo apoptosis (self-death) if that repair fails must be some of the first mutations to occur for a cell to become cancerous. Afterward, the cell must become mutated in the tumor suppressor genes (e.g., p53) that limit the proliferation rates of cells, as otherwise the detrimental effects of increased growth will not occur. Next, the cell may be mutated in the ability to proliferate such that its proliferation is increased. This may include multiple mutations as the rate of proliferation is controlled in multiple ways. Mutations in a different order will not create a cancer cell, but instead will create a cell that will undergo either repair or apoptosis to maintain the system [20, 30, 48]. An overview of the genetics of cancer progression is provided by Michor [36].

With these mutations the cell has accomplished four of Hanahan and Weinberg's requirements. Additionally, for a tumor to grow large enough to be problematic it must be able to induce angiogenesis, or the creation of new blood vessels referred to as the vasculature. New blood vessels are created due to the diffusion of the chemical VEGF by cells that are not receiving enough nutrients from the vessels ("hypoxic" cells), that alerts the vessels to the need to sprout new vessels. These new blood vessels will bring new nutrients to the cancer cells, allowing the tumor cluster to grow larger. We create a multiscale agent-based model of cancer to study the growth of tumor cells through the phase of angiogenesis, based on these hallmarks of cancer. This work is based on our previously published work [38].

3.1 Related Work

There are many valid techniques for studying cancer; the choice of spatial vs. non-spatial (such as an ODE) model can have a huge impact on the results. We choose to utilize a spatial representation of cells as it is more similar to the original biological system and thus allows for the inclusion of a greater number of detailed aspects.

To explicitly represent space the primary options are partial-differential equations (PDE), a cellular automata model, or an agent-based model. We choose to use agent-based models as they are particularly well suited for examining multiple scales. They also have the benefit of being able to be simulated in a more accurate way than how one would simulate a PDE.

This chapter does not describe the first agent-based cancer simulation. CancerSim, an earlier three-dimensional agent-based model, begins with a cancer-free system and then models the growth of the cancerous cells based on the hallmarks of cancer [1, 22]. It therefore takes into account basic cellular properties such as genetic instability, telomere length, and random apoptosis. Another three-dimensional hybrid agent-based tumor model utilizes basic gene-protein interactions and multi-cellular patterns specific to brain cancer [59]. This model represents internal cellular processes via differential equations, and the location of cells spatially. An agent-based approach has also been used to investigate the role of the immune system in reacting to and fighting tumor growth [17].

However, these models don't incorporate angiogenesis, the growth of new blood vessels to provide tumors with enough nutrients to continue to grow. Models of angiogenesis fall into three categories: models of vasculature (vessel) growth without tumor or other tissue cells, models of vasculature growth with only tumor cells, and models of vasculature growth with both tumor and healthy tissue cells. Generally, models of angiogenesis include the growth of vasculatures to support tumors and the size of tumors with and without angiogenesis. In some models angiogenesis is just one of the mechanisms modeled in order to answer a different question, such as the shape of growth or time of growth of the tumor.

Prior mathematical modeling can inform our model design, such as in terms of how capillaries are formed, the flow of blood, vessel adaptation, and the extent of which chemical diffusion reaches the tissue. McDougall et al. analyze how drugs can be sent to the tumor via the formed vessel network, based on the quality of the created vessels [35]. They found that the simulation was sensitive to changes in parameters for haptotactic response of the blood vessel cells, blood viscosity, and blood pressure, and that the tracer-drug sent through the system was able to travel easiest through the well formed vessels. A discrete mathematical model based on partial differential equations has been used to examine the spatio-temporal evolution of capillary networks in two and three dimensions, finding that they can represent a realistic structure [33].

A two-dimensional multi-scale model of angiogenesis where endothelial cell sprouting occurs due to VEGF diffused by healthy cells in an initial development phase, and by tumors that are later implanted into the model, has been developed [43]. Sprouts grow based on VEGF gradient that occurs when cells do not have enough oxygen. Sprouts die if they do not create a loop within a given period of time, and are only active once it is a viable segment. They find that a larger number of tumor cells causes a higher vascular density, and that network remodeling requires a balance between angiogenesis and vessel pruning. A two-dimensional PDE model of tumor growth with angiogenesis finds that asymmetrical tumor growth leads to a greater degree of branching at the surface of the tumor when compared to symmetrical tumor growth [10].

Multiple angiogenesis models are based on the cellular Potts model. One such model found that sprout morphology is not affected by how far from the tip the proliferating region is, although the speed of growth is affected [6]. Another Potts model analyzes tumor growth with and without angiogenesis [52]. The initial vasculature is represented as a grid. Endothelial sprouts grow in biased random movement toward VEGF secreted by hypoxic tumor cells, with the only modeled nutrient being oxygen. They find that the system goes through six phases of growth, and that their results are similar to other models. We include similar aspects in our model, although we use a different level of abstraction, and thus will compare our results to [52].

Some cellular automata models focus on branching rules and how they change the vasculature, with a brief description of how it may be applied to tumor angiogenesis [34]. Others focus specifically on tumor angiogenesis, with either a visual comparison to in vivo endothelial sprouting [45] or a comparison with data on in vivo vascular growth [46]. Both find that their sprouting shape is similar to what would be expected from tumor angiogenesis.

None of the above models consider how the cancer cells affect the rest of the system, however. We design our model to include healthy tissue cells instead of the immune cells. Including other types of cells leads to a more realistic model, as surrounding cells have already been shown to play a role in affecting cancer growth. The agent-based model described in this chapter is designed to incorporate many aspects of cancer growth including angiogenesis, but with algorithmic choices that allow a complex model to still be able to be run in only a few hours. These algorithmic choices include an improved abstraction of angiogenesis, and the use of stigmergy to facilitate cellular adhesion and vessel growth.

3.2 Modeling Approach

In our model each agent represents a single cell, and exists on a three-dimensional grid to approximate a tissue. The system is modeled on three scales (Fig. 5): molecular (VEGF, diffusion), cellular (genes, space) and tissue (vasculatures, branching). The molecular level represents the chemicals that are diffused in the system to enable self-organization. The cellular level contains details on how a single cell functions in the system, and the tissue level is the environment in which cells and chemicals coincide.

Although two dimensions can also be modeled, three dimensions provides a more realistic representation. Agents follow individual "life protocols" defined and implemented through probabilities represented within their computational genes. We consider normal tissue cells to be of size 10 μm, and each time step to represent approximately one day. Vasculatures in the system give agents nutrients, which they require to survive. Without tumor growth, the system can self-maintain essentially forever.

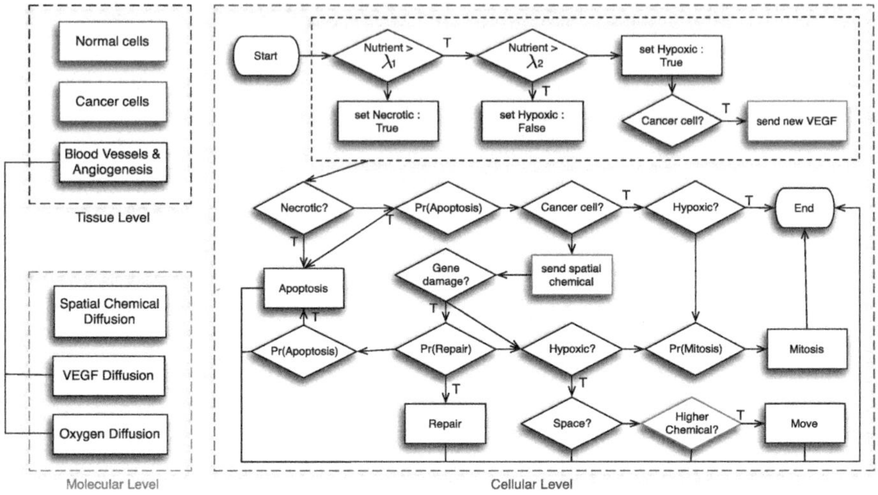

Fig. 5 Model overview with the basic relationships between the three scales, including a cellular level flow chart of agent decisions for both normal and cancer cells, for every time step. Normal and cancer cells also differ by what probabilities are used for decisions. Angiogenesis is modeled both at the tissue level (branching of new sprouts) and at the molecular level (diffusion of oxygen and VEGF)

3.2.1 Cellular Scale

The basic life protocols are based on Hanahan and Weinberg's hallmarks of cancer [22], and thus reflect proliferation (including rate parameters, generation potential, and space restrictions [2, 5, 27]), proliferation-suppression mechanisms, self-testing at a check point prior to the replication decision, repairing damage, and apoptosis (self-death). The latter is activated as a random process, secondary to an agent's decision to die due to aging or uncorrected defects. The distance regulation protocol maintains shape cohesiveness and allows undisturbed communication flow among agents [3]. A flow chart of the decisions by a cell at every time step can be seen in Fig. 5.

Tumor agents develop in the system when all of an agent's life protocols are damaged. A tumor agent can only produce tumor agents and cannot mutate back into a normal agent. Since all life protocols are broken, a tumor's ability to follow each of those protocols is also broken: it is not able to repair the mutations, it cannot undergo apoptosis that would occur due to mutations, and it proliferates more frequently than a healthy cell.

Each cell, whether or not it is a tumor cell, is in one of three states at any given time related to its nutrient levels (Fig. 6): normal, hypoxic, or necrotic. Cells are in the normal state if they have received enough nutrients, i.e. above the set threshold described in the next subsection. Normal state cells will function as described. Cells become hypoxic when they have received some nutrients, but not enough to be above the threshold to be considered normal. Hypoxic cells survive, but are unable to

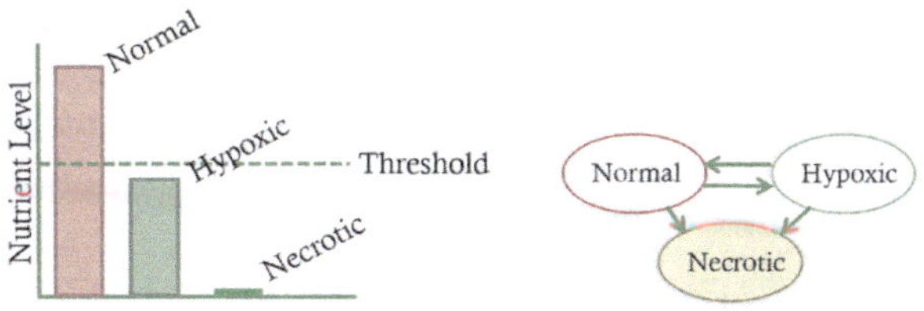

(a) Nutrient Levels (b) Cellular States

Fig. 6 Nutrient Levels and Their Effect on Cells: **a** The nutrient levels pertaining to each of the three potential cell states **b** a representation of how cells move between states

proliferate. Necrotic cells have not received any nutrient, and undergo apoptosis. The only difference for tumor cells is that hypoxic cells will diffuse the molecule VEGF in an attempt to grow new blood vessels through the process known as angiogenesis, described below.

3.2.2 Molecular Scale

There are three types of chemicals diffused within the molecular level of this model: spatial chemicals, nutrients, and VEGF. The spatial chemical is diffused by each normal cell to enable self-organization in creating a cohesive tissue by alerting other cells to its presence; it is constantly diffused and rapidly decays. This chemical allows normally functioning cells to determine in what direction the highest proportion of cells exists within their larger neighborhood; a higher level of chemical in one direction signifies more cells, and thus the center of the tissue. If a cell is moving, it will move toward the highest chemical amount to enable adhesion. This stigmergy enables normal cells to grow in a spheroid initially as they populate the system prior to any tumor cell creation, and retain shape over time such that single cells don't wander away from the overall structure. Tumor cells neither create nor use this chemical.

Each blood vessel diffuses nutrients to nearby agents in all directions, as all cells require nutrients to survive. There is a limited amount of nutrient that can be supplied by a single vessel, and a limited distance that it can travel from the vessel. Additionally, as cells use nutrient it decreases the amount of nutrient available for cells further from the vessel. Thus, vessels must occur close enough to sustain all cells in the simulation, but optimally should be as far apart as possible to accomplish that goal realistically. We model oxygen as it is generally the most limiting nutrient sent to cells.

We represent blood vessels at equidistant locations throughout the tissue (every 200 μm since oxygen can diffuse up to 100 μm in vivo), forming a grid. We further view the grid as creating a series of compartments, with each compartment being bounded on three sides by vessels, one in each of the three directions. These vessels can supply nutrients to all healthy cells within the tissue, assuming there are no tumor cells present. Nutrients are distributed to cells from each vessel, first to the cells closest to the vessel, then to the next layer of cells past that, etc, until the maximum number of cells that vessel can support has been reached in each of the three dimensions surrounding it. We assume blood flows from the zero location to the maximum location for each vessel. Each vessel also has a maximum number of cells that can be supported overall in the model, in addition to the maximum amount within each region. Thus, if a region closer to the beginning of the tissue has additional cells that use more oxygen, the cells further along the vessel will be more likely to become hypoxic.

Algorithm 1 Diffusion of nutrients

```
compartmentNutrientMax
vesselNutrientMax = compartmentNutrientMax * numberOfCompartments
for each compartment in vessel do
  availableNutrient = compartmentNutrientMax
  for each diffusion radius within diffusionRadiusMax do
    for each location from 0 to compartmentWidth at this radius do
      if availableNutrient < 0 then
        availableNutrient = 0
      end if
      send availableNutrient to each of N cells at location
      availableNutrient = availableNutrient - N
    end for
  end for
end for
```

This technique for representing vessels and their flow is unique among angiogenesis models. Other models have utilized a grid, but they rely on specific calculations at every location for the amount of nutrient available. To decrease the complexity in the model, instead of explicitly modeling the diffusion process we model the effect of diffusion. Usually diffusion is modeled using PDEs that must then be solved to determine the amount of gradient available at every location within the grid. Even discretizing this using standard grid diffusion techniques becomes computationally expensive since the amount of nutrient available to a cell not only relies on the amount of nutrient in its location, but also on how many cells have taken nutrient from the environment before the nutrient reaches that cell. To take this issue into account, we instead do a stepwise diffusion from each vessel as seen in Algorithm 1. The *compartmentNutrientMax* is defined as in Eq. 8. α represents the number of direct neighbors in *numDimensions* $- 1$ (in the case of three-dimensions, $\alpha = 8$). The *diffusionRadiusMax* is defined as *compartmentHeight*/2.

$$compartmentNutrientMax = \sum_{i}^{compartmentHeight} (\alpha * i) * compartmentLength \tag{8}$$

Cells that receive enough nutrients are considered normal and function as described previously. It is expected that cells will receive nutrients from three vessels: one in the X direction, one in the Y direction, and one in the Z direction. Thus, the minimum amount of nutrient for a cell to remain normal is three times what each vessels should give them (λ_1). Cells that receive nutrients but not enough to remain normal (λ_2) will become hypoxic, and thus unable to proliferate until the nutrients increase again. Cells can return from the hypoxic state to a normal state each time

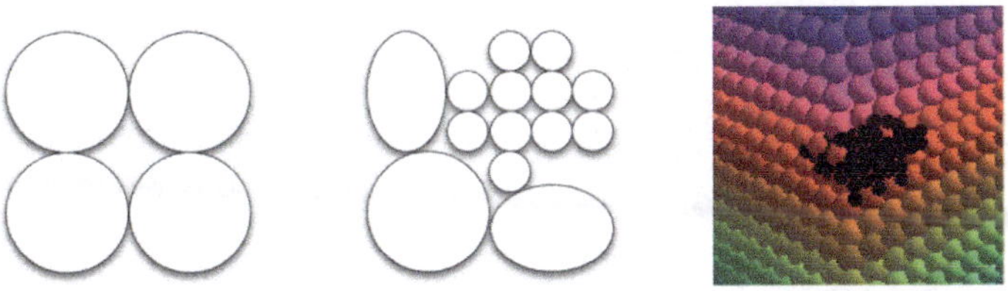

(a) Normal cells within a grid

(b) Tumor cells invading the grid

(c) Simulation

Fig. 7 Relative size of normal and tumor cells at the tissue level, in 2D for easier viewing. **a** Normal cells are by default larger, **b** but are resized as tumor cells push their way into the system. **c** In the simulation, tumor cells are shown in black and normal cells are colored. Although cells are not spherical in vivo, this abstraction allows for reasonable modeling without altering results

step, if their nutrients are $> \lambda_1$. Cells that receive zero nutrients will become necrotic and die immediately.

All hypoxic tumor cells diffuse VEGF in an attempt to induce the growth of new blood vessels to provide more nutrients to the region (angiogenesis, Fig. 5). As VEGF travels through the system, it is spread from its current location within its immediate Moore neighborhood. The amount of VEGF at the location ($currentVEGF$) is initially 1 for the hypoxic cell's (x,y,z), with 90% diffusing to neighbors. The amount diffused to an individual neighbor is therefore $0.9 * currentVEGF/(8 * numDimensions + numDimensions - 1)$, and is added to any VEGF already at that location. The diffusion continues from all locations with VEGF greater than the minimum of 0.0001, with $0.9 * currentVEGF$ diffusing among the neighbors. VEGF only disappears from the system when the amount at a particular grid point is below the minimum. Once the chemical reaches a vessel it will induce the creation of a sprout if there is not already one growing from that vessel within 30 μm. VEGF is essentially another form of stigmergy. The VEGF molecules are an indirect message from tumor cells to endothelial cells to grow new sprouts. The VEGF then also acts as an indirect form of communication pertaining to *where* the sprout cells should grow, as they can grow toward the highest concentration of VEGF. The diffusion thus serves two purposes without any additional calculations.

3.2.3 Tissue Scale

Spatial Representation

Three-dimensional space is represented by a grid. Normally functioning tissue cells generally require more space than tumor cells (Fig. 7). Our model abstracts this necessity by assuming normal cells to be twice the diameter of tumor cells. Their location is determined by a set of coordinates (x,y,z) that define their center point. With no neighboring tumor cells, the cell retains a diameter of 2 grid locations. Thus, you can consider each normal cell to be on every other lattice point.

(a) Initial vasculatures

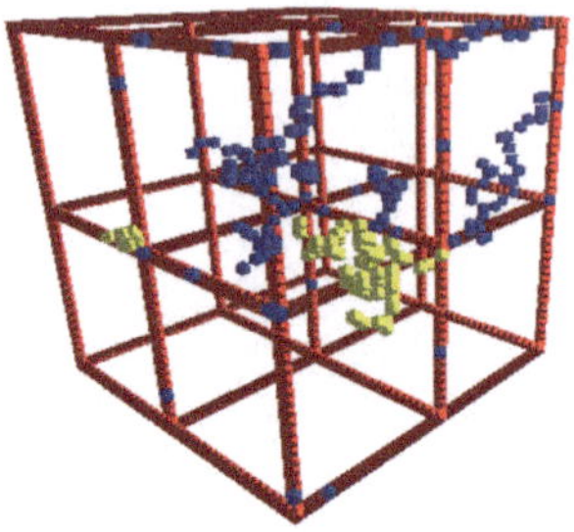

(b) Angiogenesis

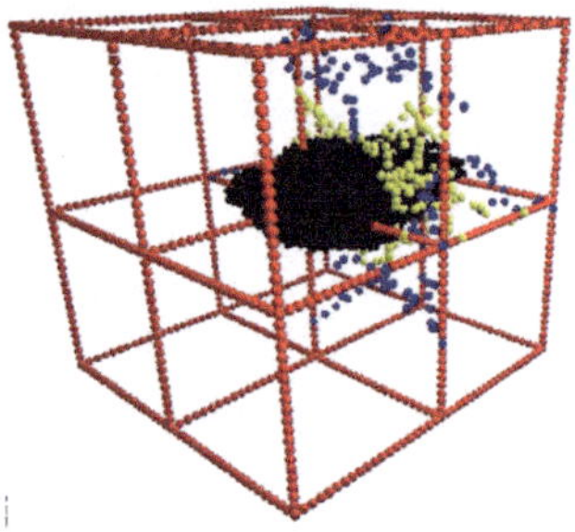

(c) Angiogenesis and tumor

Fig. 8 Modeled vasculatures. **a** The initial vessels capable of providing oxygen to the simulated tissue. **b** Vessel growth due to tumor diffusion of VEGF; yellow (light) are viable vessels that have formed a loop and thus diffuse oxygen, whereas blue (dark) are non-viable endothelial cells. **c** Vessel growth showing the tumor as well (dark center cells)

The addition of tumor cells disrupts this status quo. A tumor cell also has a coordinate (x,y,z) but only a diameter of 1 grid location. Thus, for every normal cell, there may be 27 tumor cells (Fig. 7). However, it will take some time before tumor cells proliferate in such a way that normal cells become reduced to this smaller size. A reduction in space affects a normal cell's ability to proliferate and move.

Blood vessels and sprouts created through angiogenesis also take up space in the model. Both are equivalent to the size of a tumor cell. This restriction adds both realism, and limits the number of agents in the model and thus the computational requirements.

Angiogenesis

Angiogenesis is the creation of new blood vessels due to tumor cells secreting VEGF, as described in the previous subsection. Before a new sprout can diffuse nutrients it must form a loop with another sprout or a vessel. In our model each sprout can grow by five endothelial cells every time step, by biased random movement based on the gradient of VEGF within the system (Fig. 8). They can only grow in locations where other cells are not already present. Once the sprout becomes viable, it provides nutrients to nearby cells; the amount of oxygen diffused in total by a sprout (ρ_s) is relative to its length (l) as seen in Eq. 9. The diffusion of this nutrient occurs via an

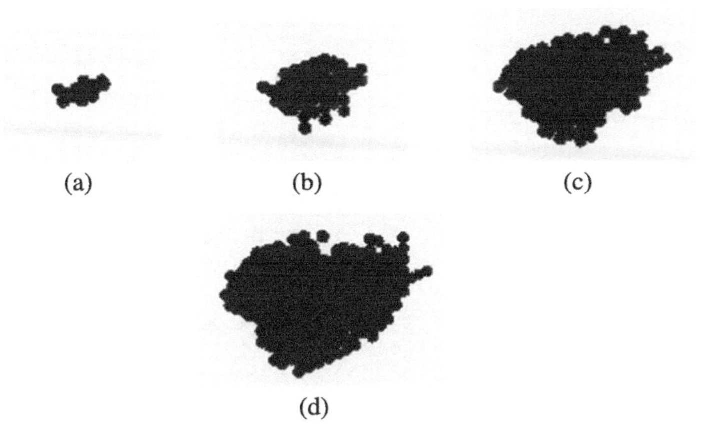

Fig. 9 General growth pattern of modeled cancer cells. Tumor exhibit spheroid growth patterns as well as the expected branching processes. Normal cells and vasculatures are not shown, although they are influencing the tumor's growth

algorithm similar to Algorithm 1, except that the diffusion radius is limited to half of a normal vessel's diffusion radius. More diffused nutrients allows further tumor growth in that region. However, the total number of nutrients in the system remains constant.

$$\rho_s = \sum_{i}^{diffusionRadiusMax/numDimensions} (\alpha * i) * l/\alpha + 5 \tag{9}$$

3.3 Results

In all instances the model is simulated on a size $40 \times 40 \times 40$ grid. Compartments have height, width, and depth of 20. Normal cells have the following parameters: proliferation rates in one of three ranges (0.05–0.15, 0.03–0.09, or 0.015–0.045), and apoptosis rates within 0.0024–0.0048. Cancer cells have the following related parameters: proliferate rates are limited to the range of 0.3–0.9, and apoptosis rates are limited to 0.0001–0.0002. These values are chosen to represent breast cancer [40]. Angiogenesis follows the above algorithms when enabled in the simulation.

Without angiogenesis, in all of the parameter cases we first see the expected quick growth of a self-sustaining cellular environment, followed by tumors proliferating at an unchecked rate once the first tumor cell is introduced, until they reach the maximum number that can be sustained. The tumor growth pattern is in the expected spheroid growth pattern although we do not explicitly model this behavior [47, 59], including expected branching (Fig. 9). Normal cells mostly die due to their lack of nutrients and inability to proliferate when surrounded by tumor cells. An example of this growth pattern can be seen in Fig. 10a.

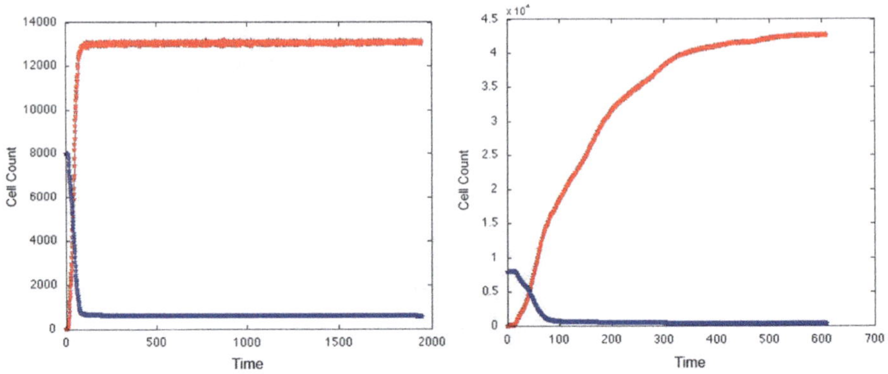

(a) Without angiogenesis (b) With angiogenesis

Fig. 10 Change over time in number of cancer (red/light) and healthy tissue (blue/dark) cells. Note the different axes. **a** When angiogenesis is not activated, cancer cells grow quickly, but are unable to grow large enough to crowd out healthy cells. **b** When angiogenesis is activated, the cancer cells are able to grow much larger and take nutrients and space away from healthy cells

As expected, cancer cells are able to grow much larger with angiogenesis, which results in normal cells essentially being removed from the system (Fig. 10b). Angiogenesis enables tumor cells to increase the nutrient supply around the tumor cell cluster, thus increasing their possible size.

We next compare the model to a previously published angiogenesis model by running the simulation with only tumor cells, as our comparison model does not contain normal cells [52]. As with the other model, we start with a single tumor cell, which eventually grows into a tumor that can no longer be sustained by the vasculature. We examine the rates of tumor cells, hypoxic cells, and neovascular cells in this model (Fig. 11) over time. The rates of tumor growth with and without angiogenesis relate as we would expect, with tumor growth increasing much faster with angiogenesis, reaching 35,000 cells in 400 days instead of 15,000 cells without angiogenesis (Fig. 11a). Additionally, the tumor grows in a spheroid pattern with small branchings, which is also expected of tumor development.

After approximately 100 days the number of hypoxic cells begins to increase (Fig. 11b). Once the vasculature can't sustain tumor growth, hypoxic cells diffuse VEGF and angiogenesis begins. The increased tumor growth rate corresponds to the increase in neovascular and hypoxic cells. We see an increase in hypoxic cells in both cases, as cells will lack nutrients whether or not the tumor cells can encourage angiogenesis. In the case of no angiogenesis ability, although the number of hypoxic cells increases the overall growth of tumor cells is delayed as the cluster can only proliferate in areas where nutrients are still available. Eventually we start to see a more exponential growth in the case of angiogenesis, showing that the tumor is able to grow but still cannot support all cells.

These results are as we expect to see, and are very similar to the results from [52]. However, the most noticeable differences between our results are the amount

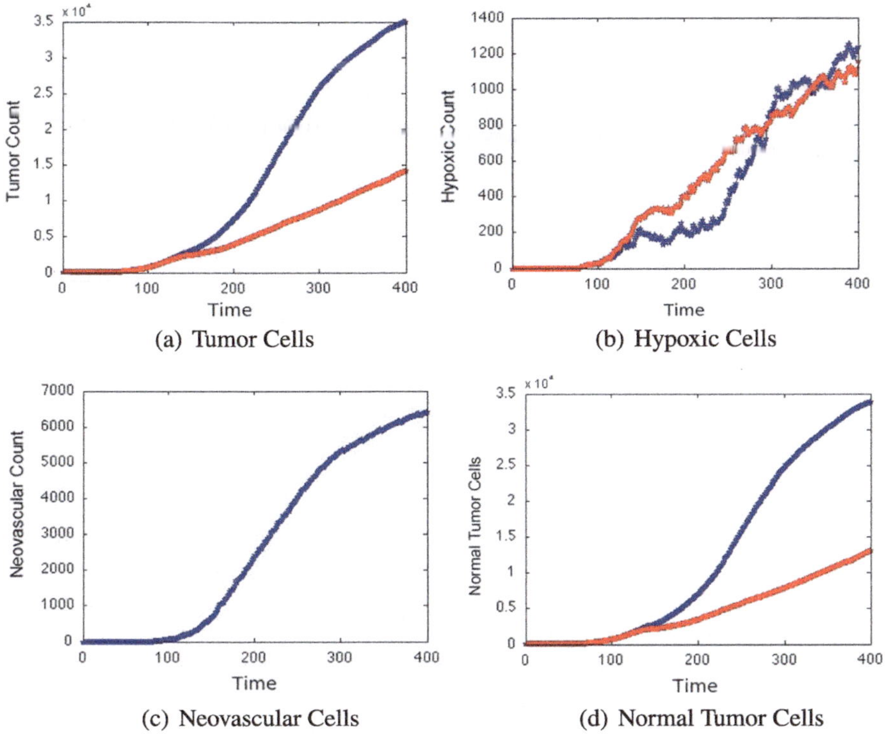

(a) Tumor Cells (b) Hypoxic Cells (c) Neovascular Cells (d) Normal Tumor Cells

Fig. 11 Change over time of cell counts with (blue/dark) and without (red/light) angiogenesis: **a** Number of tumor cells, **b** Number of hypoxic (low nutrient) cells, **c** Number of neovascular (new vessel) cells, and **d** Number of non-hypoxic tumor cells. More tumor cells are able to grow with angiogenesis, and eventually there is also more rapid creation of hypoxic cells with angiogenesis. These results resemble the first five phases in [52]

of time to see these patterns, as well as the number of tumor cells simulated. These differences may be an artifact from our inclusion of normal tissue cells, showing that to truly model cancer growth we need to include these cells in our simulations.

Stigmergy plays a key role in being able to model cancer cell growth in an agent-based model. Without diffusion of VEGF messages between tumor cells and vessels, additional growth would not be possible. In this situation direct communication does not make sense, as tumor cells are not directly connected with vessels, nor does trail communication as cells are not moving. Normal cells also rely on diffusion to understand the density of their neighborhood; diffusion in this case allows cells to spread information for any interested party in nearby locations to cause awareness of cell locations. The key for stigmergy functioning in a such a complex model with multiple types of diffusion occurring at every time step is to develop algorithms for diffusion that do not require constant recalculation of diffused chemicals.

4 Discussion

Many biological systems can be studied using a complex adaptive systems approach. In this chapter we've described two such systems: collaboration within a predator-prey system, and angiogenesis in cancer growth. The first utilized a cellular automata model, and the second a multiscale agent-based model.

In both of these situations there is some form of emergent behavior, as simple rules cause complex behavior. This behavior is accomplished by a form of stigmergy, indirect communication of information about the world to neighbors. For the predator-prey model, the internal emotional state of each entity within the system is shared indirectly with neighbors through the environment. That shared information is then incorporated into the neighbor's own internal emotional state to influence its future decisions. We tested two approaches to communication between conspecifics: a version of stigmergy mimicking a trail of ant pheromones, and a simple local communication in which only current neighbors are told about an entity's emotion and thus there is no diffusion. We find that both predator and prey benefit the most from being the only species to use emotions in their decision making. The predator population also benefits when both populations use emotion, but only if they are using trail communication with a geometric decay. Thus, our choice of communication paradigm as well as the parameters to that paradigm affect the result.

In the cancer angiogenesis model, stigmergy was used in multiple forms: a diffusion of spatial chemical to aid cellular adhesion, and a diffusion of VEGF to encourage vessel sprout growth. The indirect communication between cells correctly mimics the cancer system in the human body, and allows us to represent complex phenomena. In this case it was necessary to be strategic in the algorithms used to perform the diffusion, to enable the simulation to run in a reasonable amount of time.

This paper defined various options of communications among entities in either a cellular automata or agent-based model, and then examined the use of communication in two different types of models. Results support that stigmergy can be adapted to a variety of situations, and also that awareness of the speed of algorithmic decisions can increase its usefulness in biological modeling. These ideas can be adapted to many other modeling situations other than the classic examples of self-organization like bees or ants.

References

1. Abbott, R.: Cancersim: a computer-based simulation of hanahan and weinberg's hallmarks of cancer. Master's thesis, University of New Mexico (2002)
2. Allsopp, R.C., Vaziri, H., Patterson, C., Goldstein, S., Younglai, E.V., Futcher, A.B., Greider, C.W., Harley, C.B.: Telomere length predicts replicative capacity of human fibroblasts. Proc. Natl. Acad. Sci. U.S.A. **89**(21), 10114–10118 (1992)
3. Alexander, R.A. Anderson, A.M. Weaver, P.T.: Cummings, and Vito Quaranta. Tumor morphology and phenotypic evolution driven by selective pressure from the microenvironment. Cell **127**(5), 905–915 (2006)

4. Anderson, P.W.: More is different. Science **177**(4047) (1972)
5. Barkai, N., Shilo, B.-Z.: Variability and robustness in biomolecular systems. Mol. Cell **28**(5), 755–760 (2007)
6. Bauer, A.L., Jackson, T.L., Jiang, Y.: A cell-based model exhibiting branching and anastomosis during tumor-induced angiogenesis. Biophys. J. **92**(9), 3105–3121 (2007)
7. Blanchard, D.C., Griebel, G., Blanchard, R.J.: Conditioning and residual emotionality effects of predator stimuli: some reflections on stress and emotion. Prog. Neuro-Psychopharmacol. Biol. Psychiatr. **27**, 1177–1185 (2003)
8. Bonabeau, E., Dorigo, M., Theraulaz, G.: Swarm Intelligence: From Natural to Artificial Systems and. Oxford University Press, Oxford (1999)
9. Bonabeau, E., Dorigo, M., Theraulaz, G.: Inspiration for optimization from social insect behaviour. Nature **406**(6791), 39–42 (2000)
10. Bo Shim, E., Kwon, Y.-G., Jong Ko, H.: Computational analysis of tumor angiogenesis patterns using a two-dimensional model. Yonsei Med. J. **46**(2), 275–283 (2005)
11. Coffey, D.S.: Self-organization, complexity and chaos: the new biology for medicine. Nat. Med. **4**(8), 882–885 (1998). August
12. Couzin, I.: Collective minds. Nature **445**, (2007)
13. Couzin, I.D.: Collective cognition in animal groups. Trends Cognit. Sci. **13**, 36–43 (2009)
14. Couzin, I.D., Krause, J., Franks, N.R., Levin, S.A.: Effective leadership and decision-making in animal groups on the move. Nature **433**, 513–516 (2005)
15. de Carvalho, K.C., Tome, T.: Self-organized patterns of coexistence out of a predator-prey cellular automaton. Int. J. Mod. Phys. C **17**, 1647–1662 (2006)
16. Dewdney, A.K.: Sharks and fish wage an ecological war on the toroidal planet wa-tor. Sci. Am. (1984)
17. Dréau, D., Stanimirov, D., Carmichael, T., Hadzikadic, M.: An agent-based model of solid tumor progression. In: Proceedings of the 1st International Conference on Bioinformatics and Computational Biology, BICoB '09, pp. 187–198. Springer, Berlin, Heidelberg (2009)
18. Ekman, P.: Basic Emotions. Wiley, New York (1999)
19. Farina, F., Dennunzio, A.: A predator-prey cellular automaton with parasitic interactions and environmental effects. Fundam. Inf. **83**, 337–353 (2008)
20. Frank, S.A., Iwasa, Y., Nowak, M.A.: Patterns of cell divisions and the risk of cancer. Genetics **163**, 1527–1532 (2003). April
21. Gardner, M.: The fantastic combinations of john conway's new solitaire game 'life'. Sci. Am. **223**, 120–123 (1970)
22. Hanahan, D., Weinberg, R.A.: The hallmarks of cancer. Cell **100**, 57–70 (2000). January
23. Harrington, K., Olsen, M., Siegelmann, H.: Communicated somatic markers benefit both the individual and the species. In: Proceedings of the International Joint Conference on Neural Networks (2011)
24. Harrington, K., Olsen, M., Siegelmann, H.: Computational neuroecology of communicated somatic markers. (2012)
25. Hawick, K.A., Scogings, C.J.: A minimal spatial cellular automata for hierarchical predator-prey simulation of food chains (technical report cstn-040). Technical report, Computer Science, Massey University (2009)
26. Hogeweg, P.: Cellular automata as a paradigm for ecological modeling. Appl. Math. Comput. **27**, 81–100 (1988)
27. Hyung Don Ryoo, T.G., Steller, H.: Apoptotic cells can induce compensatory cell proliferation through the jnk and the wingless signaling pathways. Dev. Cell **7**(4), 491–501 (2004)
28. Kitano, H.: Systems biology: a brief overview. Science **295**(5560), 1662–1664 (2002)
29. Lehman, C.L., Tilman, D.: Competition in Spatial Habitats, pp. 185–203. Princeton University Press, Princeton (1997)
30. Lindahl, T., Wood, R.D.: Quality control by dna repair. Science **286**(3) (1999)
31. Lotka, A.J.: Elements of Physical Biology. Williams and Wilkins (1925)
32. Low, A., Lang, P.J., Smith, J.C., Bradley, M.M.: Both predator and prey: emotional arousal in threat and reward. Psychol. Sci. **19**, 865–873 (2008)

33. Mark A.J.C.: Mathematical modelling of angiogenesis. J. Neuro-Oncol. **50**, 37–51 (2000). https://doi.org/10.1023/A:1006446020377
34. Markus, M., Böhm, D., Schmick, M.: Simulation of vessel morphogenesis using cellular automata. Math. Biosci. **156**(1–2), 191–206 (1999)
35. McDougall, S.R., Anderson, A.R.A., Chaplain, M.A.J.: Mathematical modelling of dynamic adaptive tumour-induced angiogenesis: Clinical implications and therapeutic targeting strategies. J. Theor. Biol. **241**(3), 564–589 (2006)
36. Michor, F., Iwasa, Y., Nowak, M.A.: Dynamics of cancer progression. Nat. Rev. Cancer **4**, 197–206 (2004)
37. Millonas, M.M.: Swarms, phase transitions, and collective intelligence. SFI Stud. Sci. Complex. **17**, 417–417 (1994)
38. Olsen, M.M., Siegelmann, H.T.: Multiscale agent-based model of tumor angiogenesis. Proc. Comput. Sci. **18**, 1016–1025 (2013)
39. Olsen, M., Harrington, K., Siegelmann, H.: Conspecific emotional cooperation biases population dynamics: a cellular automata approach. Int. J. Nat. Comput. Res. **1**(3), 51–65 (2010)
40. Olsen, M.M., Siegelmann-Danieli, N., Siegelmann, H.T.: Dynamic computational model suggests that cellular citizenship is fundamental for selective tumor apoptosis. PLoS One **5**(5) (2010)
41. Olsen, M., Harrington, K., Siegelmann, H.: Computational emotions in a population dynamics cellular automata encourage collective behavior. In: International Conference on Complex Systems (2011)
42. Olsen, M.M., Fraczkowski, R.: Co-evolution in predator prey through reinforcement. J Comput Sci. **9**, 118–124 (2015). https://doi.org/10.1016/j.jocs.2015.04.044
43. Owen, M.R., Alarcon, T., Maini, P.K., Byrne, H.M.: Angiogenesis and vascular remodelling in normal and cancerous tissues. J. Math. Biol. **58**(4–5), 689–721 (2009)
44. Park, K.: The internet as a complex system. In: Park, K., Willinger, W. (eds.), The Internet as a Large-Scale Complex System, pp. 1–89. Oxford University Press, Oxford (2005)
45. PawełTopa. Dynamically reorganising vascular networks modelled using cellular automata approach. In: Proceedings of the 8th International Conference on Cellular Automata for Reseach and Industry, ACRI '08, pp. 494–49. Springer, Berlin (2008)
46. Peirce, S.M., Van Gieson, E.J., Skalak, T.C.: Multicellular simulation predicts microvascular patterning and in silico tissue assembly. FASEB J. **18**(6), 731–733 (2004)
47. Pettet, G.J., Please, C.P., Tindall, M.J., McElwain, D.L.S.: The migration of cells in multicell tumor spheroids. Bull. Math. Bio. **63**, 231–257 (2001)
48. Plotkin, J., Nowak, M.A.: Different effects of apoptosis and dna repair on tumorigenesis. J. Theor. Biol. **214**, 453–467 (2002)
49. Plutchik, R.: The nature of emotions. Am. Sci. **89**, 344–350 (2001)
50. Rolls, E.: What are emotions, Why do We Have Emotions, and What is Their Computational Basis in the Brain?. Oxford University Press, Oxford (2005)
51. Rohilla Shalizi, C.: Methods and techniques of complex systems science: An overview. In: Micheli-Tzanakou, E., Deisboeck, T.S., Yasha Kresh, J. (eds.), Complex Systems Science in Biomedicine. Topics in Biomedical Engineering International Book Series, pp. 33–114. Springer, Berlin, (2006)
52. Shirinifard, A., Scott Gens, J., Zaitlen.: 3d multi-cell simulation of tumor growth and angiogenesis. PLoS ONE **4**(10), e7190 (2009)
53. Simon, H.: The Organization of Complex Systems. George Braziller (1973)
54. Sirot, E., Touzalin, F.: Coordination and synchronization of vigilance in groups of prey: the role of collective detection and predators' preference for stragglers. Am. Nat. **173**, 47–59 (2009)
55. Sumpter, D.J.T.: The principles of collective animal behaviour. Phil. Trans. R. Soc. B **361**, 5–22 (2006)
56. Sumpter, D.J.T., Beekman, M.: From nonlinearity to optimality: pheromone trail foraging by ants. Anim. Behav. **66**, 273–280 (2003)

57. Wolkenhauer, O., Auffray, C., Baltrusch, S., Blthgen, N., Byrne, H., Cascante, M., Ciliberto, A., Dale, T., Drasdo, D., Fell, D., Ferrell, J.E., Gallahan, D., Gatenby, R., Gnther, U., Harms, B.D., Herzel, H., Junghanss, C., Kunz, M., van Leeuwen, I., Lenormand, P., Levi, F., Linnebacher, M., Lowengrub, J., Maini, P.K., Malik, A., Rateitschak, K., Sansom, O., Schfer, R., Schrrle, K., Sers, C., Schnell, S., Shibata, D., Tyson, J., Vera, J., White, M., Zhivotovsky, B., Jaster, R.: Systems biologists seek fuller integration of systems biology approaches in new cancer research programs. Cancer Res. **70**(1), 12–13 (2010)
58. Yamada, K.M., Cukierman, E.: Modeling tissue morphogenesis and cancer in 3d. Cell **130** (2007)
59. Zhang, L., Athale, C.A., Deisboeck, T.S.: Development of a three-dimensional multiscale agent-based tumor model: Simulating gene-protein interaction profiles, cell phenotypes and multicellular patterns in brain cancer. J. Theor. Biol. (2007)

Strategic Group Formation in the El Farol Bar Problem

Andrew J. Collins

Abstract The El Farol bar problem sprung from one patron's wish to avoid the bar's busy nights in the 1990s. The problem became of interest to economists because of its potential application to other consumer choice behavior problems, e.g., route selection on a congested roadway. The El Farol bar problem involves multiple decision-making agents trying to outwit each other and only attend the bar when it is not overcrowded. Each agent makes their decision based on historical data and draws their attendance strategy from a limited pool. We adapt the model of the problem developed, by Rand and Wilensky, to include group decision-making behavior and strategic group formation. In our version, agents can use the best strategy from the whole group, not just their own set. However, the larger the group, the more it adds to the overcrowding issue. Thus, an agent must balance access to a large attendance strategy pool with group size. We had hypothesized that including strategic group formation will increase overall social welfare, but our analysis shows that allowing agent groups results in a undesirable scenario for all agents; this is due to the limited rationality of the agents.

Keywords Agent-based modeling and simulation · Group formation · El Farol bar problem

1 Introduction

In a desire to only attend a local bar when it was not busy, a researcher at the Santa Fe Institute derived the El Farol bar problem game [1]. The players (agents) use strategies based on historical attendance numbers to determine whether to attend the bar or not. The game became important within the economics' community because it was a practical example using bounded rationality (i.e., the players did not know if the other players would attend on a given night) which resulted in selfish players cooperating without direct communication among them (i.e., if a player chooses to

A. J. Collins (✉)
Old Dominion University, Norfolk, VA 23529, USA
e-mail: ajcollin@odu.edu

T. Carmichael et al. (eds.), *Complex Adaptive Systems*, Understanding Complex Systems, https://doi.org/10.1007/978-3-030-20309-2_9

attend the bar one night, this is communicated to the remaining players through the attendance record). Beyond bar attendance, the game has been adapted to restaurant selection [2] and market behavior [3, 4].

We adapt the El Farol bar game to include strategic group formation, using a similar group formation mechanism we developed in previous papers [5, 6]. Agents strategically form groups to give them access to a larger strategy pool. However, too large a group will be undesirable because the group's attendance to the bar might cause it to become overcrowded. Our simulation is an adaption of El Farol bar simulation built in the NetLogo agent-based modeling environment [7].

The next section gives some background to the El Farol bar problem and strategic group formation. This discussion is followed by an overview of the model and its results. Finally, conclusions are given.

2 Background

This section intends to give more details on the El Farol problem and strategic group formation, both regarding the mechanism's details and it's placement in the wider literature. Our strategic group formation method uses a combination of cooperative game theory and agent-based modeling. The El Farol problem has a rich history of application since the nineties whereas our strategic group formation work is relatively new.

2.1 El Farol Bar Problem

El Farol bar (www.elfarolsantafe.com) is a popular evening entertainment destination in Sante Fe, New Mexico. Due to its popularity, the bar can become overcrowded making it a less enjoyable experience for its patrons. A researcher at the Santa Fe Institute, Brian Arthur, noticed that only on some nights the bar was overcrowded and there was no obvious pattern to which nights these would be. The problem of determining which nights to attend evolved into the El Farol bar problem.

The El Farol bar problem can be represented within an agent-based model; we used the NetLogo version [7]. In this model, the patrons are bounded rational agents who need to determine which nights to attend based on a limit number of strategies. The agents have bounded rationality because they do not have a complete history of attendance from which to make a decision, they do not have a complete strategy set available to them, and they do not know whether the other agents will attend the bar on the current night. The agents are assumed to have memory size "m" of past attendance numbers (which they know even if they did not attend a particular night). A strategy is defined as a (m + 1)-tuple s such that:

$$s \in [-1, 1]^{m+1}$$

$$P_t = s_1 + s^T U h_t$$

where "h_t" is the m-tuple of the last m attendance percentages at time "t". The first value of a strategy is a constant. "U" is a (m + 1) x m matrix that removes the first row of a transposed (m + 1)-tuple. "P_t" is the agent's predicted attendance for night "t"; if this is less than the overcrowding threshold (60%) then the agent chooses to attend. Note that it is possible that this value is less than zero, meaning the agent will always attend. All agents make their decision simultaneously.

The agents have "k" strategies to choose from in each round. The agents choose that strategy that gives the lowest total absolute error for the previous "m" nights. As absolute values are used, the agent equally weights attending on a pleasant night as well as not attending on an overcrowded night. The agent is likely to change their strategy, over time, in this complex adaptive environment of "n" agents (assumed to be 100).

Elements of El Farol model have their roots in economic interests of the 1990s: using bounded rationality in a large pool of agent [8] and learning in games [9]. However, the El Farol is a unique problem because of its focus on all agents trying to choose what the minority will be. In fact, the El Farol problem was generalized to a class of games called minority games [3, 4]. Minority games are useful because they have practical implications, for example, understanding the implications of drivers wishing to choose the least congested route or diners wishing to choose the least packed restaurant [2].

There have been several adaptions of original El Farol bar model. Fogel et al. [10] included a genetic algorithm for creating the agents' strategy pool. There has also been work on adapting the decision mechanism of agents, by allowing them to consider the action of their neighbors [11, 12]. In their papers, Chen and Gostoli considered different neighborhood structures when the agents were placed in a grid pattern and determined that the type of neighborhood structure has an impact on the outcome of the model (with von Neumann neighborhoods being the best). They also allowed the agents to have social preferences, that is, some agents were inequity averse. They should have even a small number of inequity averse agents can results in a good use of the bar, i.e., high attendance without overcrowding. None of these adaptions included strategic group formation.

2.2 *Strategic Group Formation*

Coalitions, or groups that have formed strategically, are found everywhere and can involve whole countries, e.g., the North Atlantic Treaty Organization (NATO), or just ad hoc alliance within a company [13, 14]. As such, many different fields of study have considered group formation, and coordination including social network analysis [15], artificial intelligence [16–18], agent-based modeling [8, 19–21], game theory [22], and hybrid modeling [21].

We are specifically interested in strategic group formation, that is, agents join a group because it provides some increase in utility to them and the group allows the agents to join because it increases the overall utility of the existing group members. Cooperative game theory is the "go to" methodology for considering strategic group formation [23]. Cooperative game theory introduces new concepts to traditional game theory, like Lloyd Shapley's, a Nobel prize winning economist, Shapley value [24]. Like traditional game theory, cooperative game theory considers the actions of multiple decision-makers who are focused on maximizing their utility. Unlike normal-form game theory, cooperative game theory provides some additional insights by adding a mechanism to capture strategic group/coalition formation. Also, cooperative game theory tends to focus on dozens of agents, as opposed to only two, hence it used to be called n-person game theory [25].

In recent year, we have been integrating cooperative game theory with agent-based modeling to try and overcome some of its computational limitations [5, 6]. The agent-based model described in this paper is an application of this work.

Agent-based modeling is more versatile than game theory because of its ability to model several different characteristics of an agent[1] in a dynamic environment. The agents in our scenario face a changing game environment as their group and groups around them change, resulting in different attendance rates to the El Farol bar. However, the agent's strategies, in agent-based modeling, are not as sophisticated as those found in game theory. Our integration approach is an attempt to bring back sophisticated agents into ABM. This is done by taking elements of one cooperative game theory solution mechanism, the core [26], and embedding them into an agent-based model.

2.3 Cooperative Game Theory

Cooperative game theory is concerned with determining which coalitions form and who cooperates with whom. The heart of cooperative game theory is the concept of an "imputation," which is the expression of the allocation of rewards among the agents. Not all imputations are the same, and some dominate others. By dominate, we mean that a subgroup of agents gains a greater total reward in one imputation than the dominated one. Obviously, that subgroup would prefer the dominating imputation to the dominated one, and it will focus on achieving it. Thus for an imputation to be satisfactory for all sub-groups, it must not be dominated by any other imputation. The group of imputations that satisfy this criterion is called the Core [26], which was popularized by Lloyd Shapley in 1960s [27]. Determining whether an imputation

[1] You could construct a game with complex agents but you would not be able to solve it. It is finding the Nash Equilibrium of a game that makes game theory limited, for example, no solution to chess, a game only involving 32 pieces and 64 grid squares, has been found to date.

should be in the Core can be problematic, as all imputations must be check for dominance. Luckily, a theorem exists for determining the Core:

Theorem: An allocation {x1, x2,..., xn} is in the Core if and only if

$$\begin{aligned} &(a) \sum_{i=1}^{n} x_i = v(N) \qquad \text{(Efficient)} \\ &(b) \forall \boldsymbol{S} : \sum_{i \in S} x_i \geq v(S) \ \text{(Subgroup power)} \end{aligned} \tag{1}$$

The first criterion of the theorem is the same as the imputation criterion (as the Core is a subset of all imputations if it exists). The second criterion ensures that no subgroup (coalition) has the bargaining power to leave the current imputation. It is this second criterion that we implement in our agent-based models. Throughout the simulation run, we are constantly looking for subgroups that would be better off if they split from their current group. In our simulation, we look for three group related behaviors:

1. Subgroup split: A subgroup determines if it would be better off without the remaining group. In the El Farol bar problem, this occurs when the subgroup contains an agent that has the best attendance strategy, in the group, and the size of the current group means overcrowding occurs when the group attends the bar. Note that agents do not mind being in a group until that group becomes so big that overcrowding occurs when they attend the bar.
2. Groups joining: If two groups attended the bar and it was not overcrowded, then the groups join if there joining would not have made the bar over-crowded on a previous night either group attended. By joining, the agents have access to a larger pool of strategies.
3. Individual leave: This is just a variant on the subgroup split mechanic but where the subgroup has only one individual in it.

These behaviors were checked every time-step in the agent-based simulation (ABS).

2.4 *Agent-Based Modeling and Simulation (ABMS)*

Though group formation exists within extant ABMS, like the famous Boids model [16] or our pedestrian evacuation models [28–30], they do not include any strategic behavior. Instead of strategic behavior, researchers have relied on probabilistic models [31] or preexisting group structures. Other non-strategic group formation mechanism in ABMS include Martínez-Miranda and Pavón [19], Celik et al. [20], Jiang et al. [21].

What differentiates our approach from other fields, like social network analysis [15], is that we are looking at dynamic strategic group formation behavior. We use the term strategic to mean that each agent makes decisions to maximize its utility.

"Dynamic" here indicates how groups change over time, as opposed to internal group dynamics that capture how groups function as a team. Our agents still have bounded rationality because they only have a limited strategy pool to choose from and they do not know what action the other players will do at the current time-step.

3 Model

Our model is adapted from El Farol bar model presented in Chapter three of Wilensky and Rand [32]. In this section, we discuss the adaption of the El Farol bar model and more discussion on strategic group formation.

3.1 El Farol Model

Except for including group formation, we make very few adaptions to Wilensky and Rand's El Farol bar model. The first adaption is that agents can choose strategies from their whole group's pool of strategies instead of just their own. This results in the entire group using a single best strategy. The second adaption is that agents score their performance with a return function. All other elements of the original model are kept the same. We use 100 agents and an overcrowding threshold of 60 (%). The memory size of the agents was fixed to five time periods with ten random strategies each.

We define the return function of the game as the exponentially smoothed average of good nights versus bad nights for the agent. A good night is when either the agent attended the bar when it was not overcrowded, or when the agent avoids an overcrowded night. A bad night is the opposite. Agents get a reward of one for a good night and zero for a bad night. The return function at time 't' for a given agent resolves to:

$$R_t = (1 - \lambda)R_{t-1} + \lambda(I_{attendence}(t) + I_{overcrowded}(t) - I_{attendence}(t) \cdot I_{overcrowded}(t))$$

The indicator functions are used to show attendance of the agent and if the bar was overcrowded; "R_0" is set to 0. The return function is used on our results and has no bearing on the mechanisms within the simulation.

3.2 Strategic Group Formation Model

In this section, we defined the different group formation/separation mechanisms used in the model. For group formation to be meaningful, there has to be a reason join a group and a reason to leave. If there is no reason to join a group, then agents will all remain in their singleton sets. If there is no reason to leave a group, then the simulation is likely to end up with just one group of all agents (grand coalition) which could result in a dissatisfactory outcome (i.e., the group never attends the bar because they, by definition of the grand coalition, always overcrowded when they do). The benefit for an agent joining a group is that they get to use the best strategy of the group. This may result in a single individual acting as a maven for the whole group [33]. The downside of being in a group is that if the group gets too large, its attendance will cause the overcrowding thus resulting in the group will always be attending on an overcrowded night.

We give a prose description of the algorithms used in the model to determine the group splitting/joining mechanisms. These algorithms are invoked by individual groups.

Subgroup split: Select a random group (without replacement). If the group did not attend an overcrowded night in the known history, then stop (there is no reason to split). If it did attend a night with overcrowding, then determine the smallest amount 'x' that bar was overcrowded when the group attended. If 'x' is greater than the group size 'N', then stop (as a subgroup leaving will have no effect on overcrowding). Randomly create a subgroup size equal to or less than 'N − x,' that contains agent who has a "best strategy," that is, a strategy which has the lowest predictive error out of all the agents' strategies in the group. Split this group from the main group.

The created subgroup has access to the "best strategy" so not disadvantaged with a lesser strategy. However, using this strategy, the subgroup has fewer members so, hopefully, the new group, will experience less overcrowding in the future (due to its smaller group size). Hence the subgroup is better off, in the short-term, by splitting and the subgroup is exercising its subgroup power. We only allow each group to split once per time-step.

Individual agents that do not have the best strategy will only leave if they are in a subgroup with an agent with the best strategy. If they left with a subgroup that does not have the best strategy, they would, myopically, be worst off as predictive power of the new subgroup would not be as good as the old group. Hence, only subgroups with an agent that has the maximum strategy need be considered.

Groups joining: Determine if there are any two groups that attended the bar and check that neither group would have experienced overcrowding in any of the previous nights they attended, that was under-crowded, due to any increase in group size if the groups had joined. If the two groups satisfy these condition, then they join. The new group is not considered again for joining in this time-step. Repeat this process if more eligible group pairs that have not joined. We only allow a maximum of one joining per group per time-step to stop mass amalgamation occurring all at once. Each attending group will be checked at least once for joining.

The logic behind this process was that the joining groups must have attended the bar at the same time to make contact to join. Secondly, if either group would have experienced crowding on an under crowded night they attended in the past, then they will not join. That is, each group wants to maintain attending the bar on nights it was undercrowded and will not increase its size to make those night overcrowded. On previous nights that the groups attended that was overcrowded, it does not matter if they join or not, the bar remains overcrowded on that night either way. The bonus to the groups for joining is that they have access to a larger strategy pool. The history of the larger group is used in future rounds.

Individual leave decision: This is similar to subgroup split so is accounted for there.

Any new groups formed will determine the best strategy to follow from all the agents formed. Hence at every time step, a test is conducted to see if the groups will join, split or stay the same.

4 Results

The simulation was run 100 times for both the case when groups were allowed and the original El Farol case when they were not. For each run, the simulation was allowed to run for 5,000 time-steps and the average return of the agent, average attendance from all the time-steps and number of splits/groups where recorded. A screenshot from one of the runs is given in Fig. 1; which shows a case when three groups have formed, each group with its own color for the agents. Some basic descriptive statistics are given in Table 1.

Our analysis focuses on comparing the two cases: groups allowed, groups not allowed (the original El Farol bar model). Since the results in Table 1 imply a large difference between two cases, we will first check the statistical significance of the

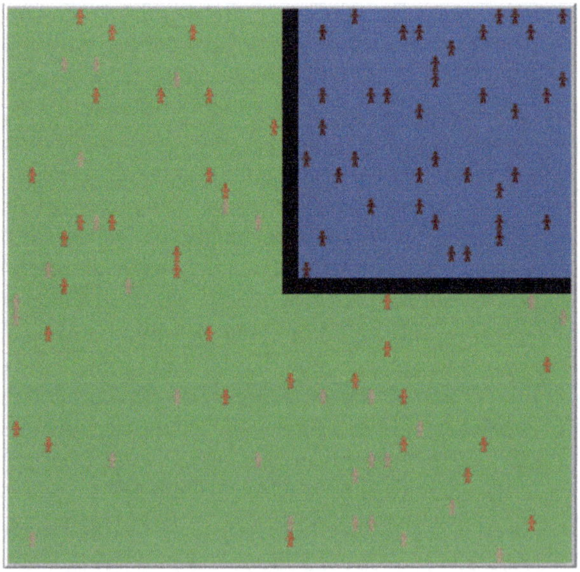

Fig. 1 Screenshot example of single group attending the bar (blue area) and remaining two groups not attending

Table 1 Descriptive statistics from Simulation Runs with each completing 5,000 time-steps

Groups allowed	Av. return	Av. attendance	Av. group size	No. groups	Group splits	Group joins
Average						
False	0.449	57.4	0	0	0	0
True	0.119	55.3	20.4	7.6	2413.4	2505.9
Standard deviation						
False	0.040	1.31	0	0	0	0
True	0.051	1.65	12.7	6.5	979.0	977.0

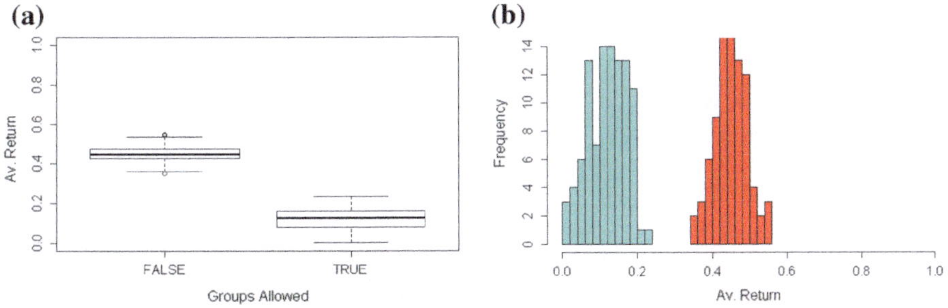

Fig. 2 Graphs comparing the average reward of the cases when group formation is allowed and when it is not. **a** A box-plot and **b** a histogram with groups allowed (blue) and no groups (red)

results before leading into a discussion into why group formation seems to hinder, not help, the agents.

Figure 2a shows a boxplot of the average return experienced by the agents. There is no overlap between the two cases with the original no-group case experiencing a much higher average reward per agent then the new group formation case. This result, of the two return, means being very different, is verified through a Welch Two-sample T-test that produces a test statistics of -50.9 and a p-value $< 2.2\text{e-}16$ which implies the result is significant at all confidence levels. The average attendance was also different between the two cases with a higher average attendance number of when groups were not allowed, which is shown in Fig. 3. The Welch Two-sample T-test that produces a test statistics of -11.3 and a p-value $< 2.2\text{e-}16$ which implies the result is significant at all confidence levels. This result occurs because when groups are allowed, there is a much higher fluctuation of in attendance from day-to-day with many days having zero attendances. This is discussed in detail in the example below.

Since a fixed number of time-steps was run, it is questionable whether convergence was reached. For the group formation case, convergence can be tested by looking to see if any group splits or joins occurred. Figure 4 shows the number of group joins over the 5,000 time-steps for each of the 100 runs. Stability is reached when no further split occurs for at least twice fixed memory of the agents (because agents would enter a fixed cycle of strategy selection and attendance selection). The graph

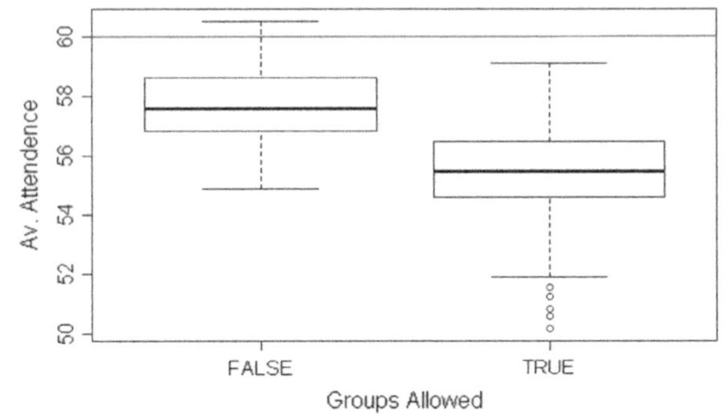

Fig. 3 Boxplot of average attendance in the cases when group formation is allowed and when it is not

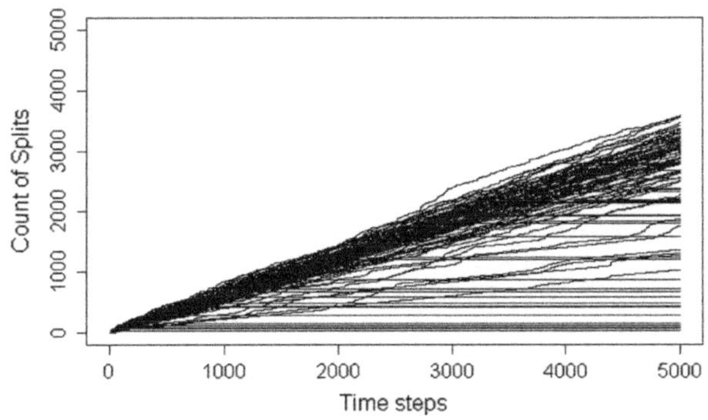

Fig. 4 Graph showing the cumulative number of group-splits for the 100 runs when group formation is allowed

implies that approximately half the runs converged because their number of splits flat-lined.

The results are initially surprising because by allowing agents access to better strategies (though a group), you might expect them to do better overall. To understand what is going on, we consider a single example. In this example, the runs have converged with all agents being part of three groups and attending in a cyclic pattern (period 12) shown in Table 2. Their groupings and strategies result in a very low average reward, which is surprising. To understand why they have converged to this results requires understanding how they would break anyway from the equilibrium and why the groups do not. There are three mechanisms in the game that would allow any group to break the equilibrium, namely: a group join, a group split, and a change in best strategy. We will discuss each in turn.

A group join would only occur if two groups attended a night that was not overcrowded. Since all nights that the groups attend is overcrowded expect when they attend alone, no joins can occur. A group split will only occur if the size of the splitting group will reduce one overcrowded night to an under-crowded night. Since all groups attend every overcrowded night, attendance at an overcrowded night is 100; thus, for a subgroup split to make a difference, 41 agents would have to be left behind to bring the number of agents attending to under the threshold of 60. Since not group size is greater than 41, this is not possible.

Table 2 Example of realized cyclic strategy after approximately 5,000 weeks. Ones indicate true and zeros false

		Weekly attendance												
Group	Size	1	2	3	4	5	6	7	8	9	10	11	12	Return
1	37	1	1	0	0	1	0	1	1	1	0	1	0	8.33%
2	26	1	1	0	0	1	0	1	1	0	0	1	1	8.33%
3	37	1	1	0	0	1	0	1	1	0	0	1	0	0.00%
Overcrowded		1	1	0	0	1	0	1	1	0	0	1	0	

The final mechanism that could create change is the agents choosing a different best strategy to follow. In our example, a change in strategy happens a lot for all three groups. However, these changes end up in a cyclic pattern resulting in the 12-day cycle that we see above. The reason these cycles occur is due to the limited memory of the agents and the limited number of strategies open to them.

To understand what is happening, let us consider group three decision at time-step 10. In the previous five days (of memory), the group has chosen very poorly and has attended every overcrowded night and not attended when the bar is empty; we represent their attendance by $(1, 0, 1, 1, 0)$. As such, the group will want to change its strategy to any strategy that is not the current one. The new best strategy would be on that had the group $(0, 1, 0, 0, 0)$ based on the past five nights, so let's pretend that this strategy is chosen. In the meantime, the other groups are also changing their strategy, so all the groups end up deciding to not attend on time-step 10, making the new best strategy a poor choice for that night. Now let us consider time-step 11, the new best strategy would have the group attend, but this is also a bad choice. Similarly with time-step 12, which results in the group wishing to change the "new" best strategy to another one and the cycle repeats.

This situation is especially complex because the groups do not have access to the optimal strategy, only the ones in their strategy set, so will not always be able to best respond to their current state. Hence, the cycle occurs because of the limit rationality of the groups (only looking five days previously) and the myopic nature of the decision making process. If there were a larger number of groups, it would be more difficult for the groups to get into a negative cycle like this. Hence the greed of the agents, wanting access to more strategies, results in their downfall.

The average return for all agents is so low for all the runs which imply that positive cycles do not occur. To understand why this is the case, consider a positive cycle, where approximately 59 agents attend each night, so the bar is under-crowded. This under-crowding would produce the maximum social welfare, across all agents, in terms of bar attendance. This requires some groups to stay at home on some nights. Now since those nights that the agents stay at home are under-crowded, those groups that stay at home current best strategy would perform badly on that night. Hence a different strategy, which allows the group to attend that night, could perform better and thus get selected by the group. This new best strategy would cause overcrowding on that given night. This process can be repeated for all the under-crowded nights, resulting in total overcrowding. This problem is particularly bad when you have a small number of large groups, each of which could overcrowd the bar if they attend.

One adaption we could make to the model to overcome this problem is to have groups consider their size when evaluating their strategies, which is, "would our attendance on an under-crowded night results in its overcrowding?" This adaption has been left to future work. Other adaptions include weighting how overcrowded the bar is on a given night when evaluating strategies.

5 Conclusions

In this paper, we adapted a version of El Farol Bar model to include strategic group formation. Patrons of the bar were able to form groups so that they had access to a larger number of attendance strategies. It was expected that groupings would result in an overall increase in return for the patrons, but the opposite was the case. The reason this happened was that each agent in a groups merged into a single decision-maker thus reducing the number of decision makers in the system which led to unhealthy behavioral patterns that emerged due to limited numbers of decision-makers. The conclusion from this work is that individuals should not form groups when wishing to be in the minority.

References

1. Arthur, W.B.: Bounded rationality and inductive behavior (the El Farol problem). Am. Econ. Rev. **84**, 406–411 (1994)
2. Chakrabarti, A.S., Chakrabarti, B.K., Chatterjee, A., Mitra, M.: The Kolkata Paise Restaurant problem and resource utilization. Phys. A: Stat. Mech. Its Appl. **388**, 2420–2426 (2009)
3. Challet, D., Zhang, Y.-C.: Emergence of cooperation and organization in an evolutionary game. Phys. A: Stat. Mech. Its Appl. **246**, 407–418 (1997)
4. Challet, D., Marsili, M., Zhang, Y.-C.: Minority games: interacting agents in financial markets. OUP Catalogue (2013)
5. Collins, A.J., Frydenlund, E.: Agent-based modeling and strategic group formation: a refugee case study. In: Proceedings of the 2016 Winter Simulation Conference, Washington, DC, pp. 1–12 (2016)
6. Collins, A.J., Frydenlund, E.: Strategic group formation in agent-based simulation. In: 2016 Spring Simulation Multi-conference, Pasadena, CA, pp. 1–8 (2016)
7. Wilensky, U.: Netlogo (1999). http://ccl.northwestern.edu/netlogo/
8. Axelrod, R.: The Complexity of Cooperation: Agent-Based Models of Competition and Collaboration. Princeton University Press, Princeton (1997)
9. Fudenberg, D., Levine, D.K.: The Theory of Learning in Games. MIT Press, London (1998)
10. Fogel, D.B., Chellapilla, K., Angeline, P.J.: Inductive reasoning and bounded rationality reconsidered. IEEE Trans. Evol. Comput. **3**, 142–146 (1999)
11. Chen, S.-H., Gostoli, U.: On the complexity of the El Farol Bar game: a sensitivity analysis. Evol. Intel. **9**, 113–123 (2016)
12. Chen, S.-H., Gostoli, U.: Coordination in the El Farol bar problem: the role of social preferences and social networks. J. Econ. Interac. Coord. **12**, 59–93 (2017)

13. Corgnet, B.: Team formation and self-serving biases. J. Econ. Manag. Strat. **19**, 117–135 (2010)
14. Wi, H., Oh, S., Mun, J., Jung, M.: A team formation model based on knowledge and collaboration. Expert Syst. Appl. **36**, 9121–9134 (2009)
15. Watts, D.J.: The "new" science of networks. Annu. Rev. Sociol. **30**, 243–270 (2004)
16. Reynolds, C.W.: Flocks, herds and schools: a distributed behavioral model. In: presented at the 14th Annual Conference on Computer Graphics and Interactive Techniques (1987)
17. Dash, R.K., Jennings, N.R., Parkes, D.C.: Computational-mechanism design: a call to arms. IEEE Intell. Syst. **18**, 40–47 (2003)
18. Cao, Y., Yu, W., Ren, W., Chen, G.: An overview of recent progress in the study of distributed multi-agent coordination. IEEE Trans. Industr. Inf. **9**, 427–438 (2013)
19. Martínez-Miranda, J., Pavón, J.: Modeling the influence of trust on work team performance. Simulation **88**, 408–436 (2011)
20. Celik, N., Xi, H., Xu, D., Son, Y.-J., Lemaire, R., Provan, K.: Simulation-based workforce assignment considering position in a social network. Simulation **88**, 72–96 (2011)
21. Jiang, G., Hu, B., Wang, Y.: Agent-based simulation approach to understanding the interaction between employee behavior and dynamic tasks. Simulation **87**, 407–422 (2010)
22. Rahwan, T., Michalak, T.P., Wooldridge, M., Jennings, N.R.: Coalition structure generation: a survey. Artif. Intell. **229**, 139–174 (2015)
23. Chakravarty, S.R., Mitra, M., Sarkar, P.: A Course on Cooperative Game Theory. Cambridge University Press (2015)
24. Shapley, L.: A value of n-person games. In: Kuhn, H.W., Tucker, A.W. (eds.) Contributions to the Theory of Games, vol. II, pp. 307–317. Princeton University Press, Princeton (1953)
25. Thomas, L.C.: Games, Theory and Applications. Dover Publications, Mineola, NY (2003)
26. Gillies, D.B.: Solutions to general non-zero-sum games. Contrib. Theory Games **4**, 47–85 (1959)
27. Shapley, L.S.: On balanced sets and cores. Nav. Res. Logist. Q. **14**, 453–460 (1967)
28. Elzie, T., Frydenlund, E., Collins, A.J., Robinson, R.M.: Conceptualizing intragroup and intergroup dynamics within a controlled crowd evacuation. J. Emerg. Manag. **13**, 109–120 (2014)
29. Collins, A.J., Elzie, T., Frydenlund, E., Robinson, R.M.: Do groups matter? an agent-based modeling approach to pedestrian egress. Transp. Res. Procedia **2**, 430–435 (22–24 Oct 2014)
30. Frydenlund, E., Collins, A.J., Elzie, T., Robinson, R.M.: Group dynamics and exit-blocking behaviors: a look at pedestrian modeling evacuations. In: 94th Transportation Research Board Annual Meeting, Washington, DC (2015)
31. Guimera, R., Uzzi, B., Spiro, J., Amaral, L.A.N.: Team assembly mechanisms determine collaboration network structure and team performance. Science **308**, 697–702 (2005)
32. Wilensky, U., Rand, W.: An Introduction to Agent-Based Modeling: Modeling Natural, Social, and Engineered Complex Systems with NetLogo. MIT Press (2015)
33. Gladwell, M.: The Tipping Point: How Little Things Can Make a Big Difference. Back Bay Books (2002)

swarmFSTaxis: Borrowing a Swarm Communication Mechanism from Fireflies and Slime Mold

Joshua Cherian Varughese, Daniel Moser, Ronald Thenius, Franz Wotawa and Thomas Schmickl

Abstract One main motivation for studying swarm intelligence comes from observing the resilience of living systems in nature. Swarm intelligence has provided important inspirations for the engineering of technical systems. The swarmtaxis algorithm and the FSTaxis algorithm are swarm intelligent algorithms that aim to move a group of agents from a starting point to a predefined goal. The swarmtaxis algorithm bases its state transition on a voting like mechanism in which the agents count the number of "pings" they get from their surroundings. In contrast, the FSTaxis algorithm uses a scroll wave based communication mechanism inspired by slime mold and fireflies. The scroll wave based communication is expected to be more resilient than the voting like mechanism of the swarmtaxis algorithm. In this paper, we borrow the communication mechanism used in FSTaxis algorithm to improve the swarmtaxis algorithm. We will also discuss how this modified algorithm performs in comparison to the parent algorithm.

Keywords Swarm intelligence · Swarm robotics · Bio-inspiration · Signaling · Taxis

1 Introduction

Nature has been a perpetual source of inspiration in scientific research and engineering. Swarm intelligence is one of the fields that relies heavily on nature as its main source of inspiration. Task allocation [11], flocking [3], navigation [2], communication [8] are few examples of the areas swarm intelligence has drawn inspiration from nature. The FSTaxis algorithm [8] implements a combination of the communication mechanisms of fireflies and slime mold for gradient ascent of a group of mobile autonomous agents. In this paper we present an improved algorithm, borrow-

J. C. Varughese (✉) · D. Moser · R. Thenius · T. Schmickl
Institut für Zoologie, Karl Franzens Universität Graz, Graz, Austria
e-mail: joshuacherian.v@gmail.com

J. C. Varughese · F. Wotawa
Institut für Software Technologie, Technische Universität Graz, Graz, Austria

T. Carmichael et al. (eds.), *Complex Adaptive Systems*, Understanding Complex Systems, https://doi.org/10.1007/978-3-030-20309-2_10

ing from this communication mechanism used by fireflies and slime mold as well as an existing swarm intelligent algorithm known as swarmtaxis [1]. This paper intends to demonstrate how resilient communication methods such as that of slime mold and fireflies can substantially improve existing algorithms.

This study is part of project subCULTron [6, 7] which aims to develop a swarm of autonomous underwater robots, which perform measurements and monitoring of the underwater environment. In contrast to traditional underwater communication using acoustics, project subCULTron uses a local communication method, blue-light communications [7], where the robots exchange small packets of modulated blue-light signals. The range of blue-light communication is around one meter under water [7]. Considering the cost and complexity of underwater swarm communication, the swarmtaxis and the FSTaxis algorithms are suitable candidates among other considered algorithms [2, 4, 9] for swarm navigation. During our recent study on resilience (which is under review for publication), we established that the communication mechanism of both FSTaxis and swarmtaxis algorithm have non-zero resilience to agent-to-agent communication failures while the FSTaxis algorithm exhibited a significantly higher resilience. In this paper, we present how the resilient communication behavior of the FSTaxis algorithm can improve the performance of the swarmtaxis algorithm.

In the Sect. 2, we will describe the algorithms that we draw inspiration from. Thereafter in Sect. 3, we will describe our approach by which we test our algorithm in simulation and present the simulation results. Subsequently, discuss the results and briefly compare it with the original swarmtaxis algorithm in Sect. 4 before concluding the paper.

2 Algorithms

In this section, we will briefly describe the swarmtaxis algorithm and then define the modified swarmtaxis algorithm.

2.1 The swarmtaxis Algorithm

In [1], the authors present the algorithm to move a group of robots from a starting point to a predefined goal. Each agent executing the swarmtaxis algorithm is required to have the ability to move, communicate with other robots within a communication range, sense other robots in its immediate surroundings and also the distance to a nearby robot. Apart from these capabilities, each robot is equipped with a long range sensor with which the robot perceive the direction of the goal. In the swarmtaxis scenario, the goal can be occluded from an agent if another agent positions itself in between the goal and the agent. If an agent is occluded from the goal, it is said to be "shadowed" and otherwise the agents are said to be "illuminated". Each agent broadcasts a single bit ping to its surroundings and every agent receiving this ping will use a polling mechanism to add it to the number of pings previously received. We will refer to this mechanism as a poll based mechanism henceforth. Depending on the pings received, the agents may assume one of the following states: "forward", "coherence", "avoid" and "random". By default, all agents are in "forward" state in which the agents moves forward at a preset velocity. If the number of pings received increases beyond a preset threshold α, the agents senses that it is at the center of the swarm and executes a random turn and then transitions back into the "forward" state. If an agent detects the number of pings it received to be lower than α, then the agent detects that it is moving away from the swarm. In this case, the agent transitions into "coherence" state in which it takes a U-turn and transitions back into the "forward" state. The states of the algorithm explained till now will only keep the agents together and not produce any net movement towards the goal. The net movement towards the goal is caused by the "avoid" states. When the robot detects another robot within its "avoidance radius", it transitions into the avoid state in which the agent takes a turn away from the detected neighbor. All the illuminated robots sets its avoidance radius slightly higher than the shadowed robots, i.e, $avoid_{illum} > avoid_{shadow}$. This enables the illuminated agents to see the shadowed agents before the latter sees the former and move away from the shadowed agents. Thus, illuminated agents moving away from the shadowed agents (but towards the goal) and in effect produces a net movement of the swarm towards the goal. As suggested in [1], α was taken to be the size of the swarm. For the sake of ease of reading, we will henceforth refer to the swarmtaxis algorithm as the "parent" algorithm.

2.2 The swarmFSTaxis Algorithm

In the swarmFSTaxis algorithm, additional to all capabilities described in Sect. 2.1, each agent is assumed to have internal timers and local directional communication. The goal can be occluded from an agent similar to the case of the swarmtaxis scenario. The behavior of the agents differ depending on whether they are "illuminated" or "shadowed". Figures 1 and 3 shows the state machine of the swarmFSTaxis algorithm.

There are two types of behaviors in the swarmFSTaxis algorithm: the "ping" behavior and the "motion" behavior just as in the case of the FSTaxis algorithm [8]. The ping behavior describes the agent-to-agent communication during the execution of the algorithm. In ping behavior, as shown in Fig. 1, the agents may assume three states: "pinging", "refractory" and "inactive". Initially, all agents are set to inactive state. In the inactive state, the agent monitors its receivers for incoming single bit local communication (pings). In the event of an incoming ping the agent broadcasts a ping for a certain duration t_{ping}. At the end of a ping, the agent enters the refractory state. During refractory time, t_{refrac}, the agent is insensitive to all incoming pings. At the end of the refractory time, the agent transitions back to inactive state. Apart from the above ping behavior, the "illuminated" agents have internal timers that are constantly counting down. When the timer counts down to zero, the agent broadcasts

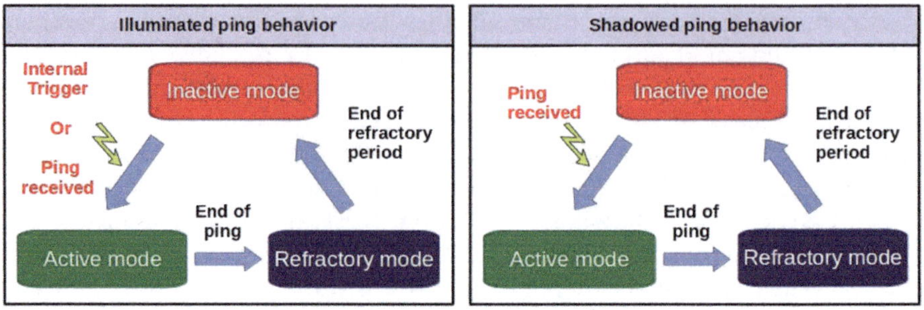

Fig. 1 A state transition diagram of the ping behavior of the swarmFSTaxis algorithm is shown in the figure. Illuminated and shadowed agents differ in ping behavior in that the shadowed agents only relay pings while the illuminated agents both produce and relay pings

Fig. 2 The ping mechanism of the swarmFSTaxis algorithm is illustrated in the figure. The "illuminated" agents has a low internal timer value and hence will hijack the pinging frequency of the system by pinging frequently. The shadowed agents will keep relaying the pings produced by the illuminated agents

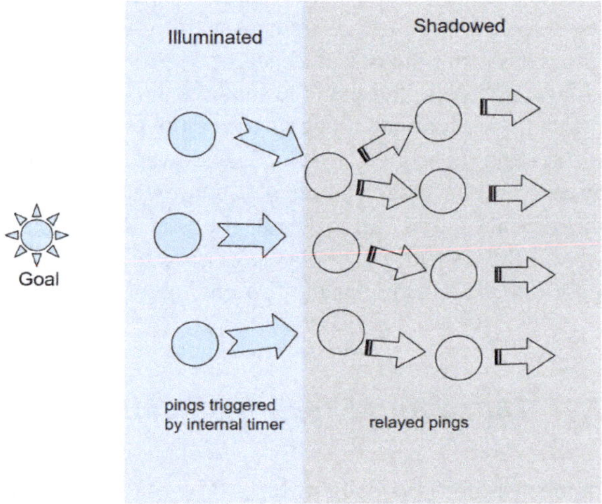

a ping. This means that an "illuminated" agent produces a ping either when the agent receives another ping or when its internal timer counts down to zero. The difference between the ping behaviors of shadowed and illuminated agents are illustrated in Fig. 2.

In addition to the ping behavior described above, the agents also have a "motion" behavior. Unlike the ping behavior, the motion behavior is same for all agents regardless of whether they are "illuminated" or "shadowed". There are two kinds of motion behavior: "general motion behavior" and "avoid motion behavior" as shown in Fig. 3. In "General motion behavior", an agent at the event of an incoming ping, moves towards the incoming ping. In case there are multiple incoming pings, the agent moves towards the mean of the directions of all incoming pings.

During "avoid motion behavior", an agent move away from a detected neighbor. As in the case of the parent algorithm, the swarmFSTaxis algorithm also imple-

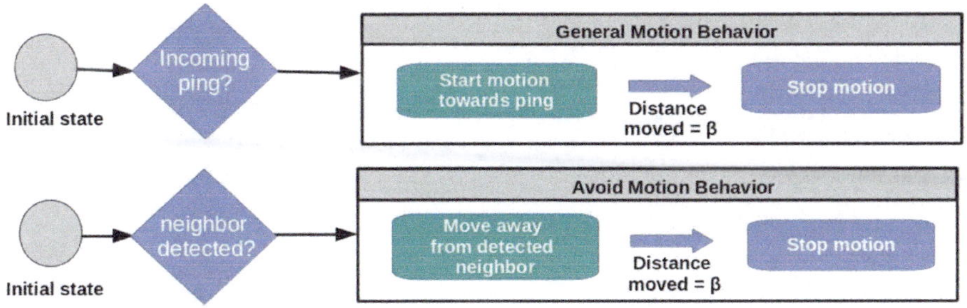

Fig. 3 A state transition diagram of the motion behavior of the swarmFSTaxis algorithm is shown in the figure. Illuminated and shadowed agents have the same motion behavior triggered by either an incoming ping or a neighbor

ments dissimilar avoidance radii for "illuminated" and "shadowed" agents: $avoid_{illum}$ and $avoid_{shadow}$. The sensor range of the illuminated agents are set to a higher value which in effect, enables the agents to move away from an approaching shadowed agent. The pseudo-code for the swarmFSTaxis algorithm can be found in Algorithm 1.

The above explained behavior will result in "scroll waves" like in the case of slime mold [5] propagating through the swarm. Since the illuminated agents trigger pings when their internal timer counts down to zero, the waves will originate at the "illuminated" agents and propagate through the "shadowed" agents as they relay the pings. The "general motion behavior" ensures that the swarm stays together with the agents moving towards incoming ping while the "avoid motion behavior" ensures that the illuminated agents move away from the shadowed agents and in effect, move towards the goal.

3 Methods and Results

The simulation environment used for testing the algorithms is Netlogo 4.3.1. In Netlogo, the simulation arena is divided into spatial units known as "patches", the simulation time is measured in "ticks". The sensor radius, s_r, of the agents is measured in coordinate distances measured from the position of the agent of interest. It is taken to be 2.5 patches since this corresponds to a reasonable underwater local communication range using blue-light in project subCULTon [7]. At the beginning of a typical run, the agents are distributed uniformly around a starting point and then the algorithm is executed. The center of mass of the swarm is used as a collective position estimate of the swarm. Once the center of mass of the swarm reaches the goal as shown in Fig. 4, the run is terminated. During the entire run, the position of the center of mass of the swarm is tracked in order to produce a representative

Algorithm 1 The swarmFSTaxis algorithm

```
repeat
   for all agents do
      PING BEHAVIOR(t_ping, t_refrac, counter, main_counter)
      AVOID MOTION BEHAVIOR(move agent, heading of detected agent)
   end for
until forever
procedure PING BEHAVIOR(t_ping, t_refrac, counter, main_counter)
   set main_counter ← main_counter + 1
   if pingmode ! = "inactive" then
      set counter ← counter + 1
   end if
   if pingmode = "active" AND counter = t_ping then
      set pingmode ← "refractory"
   end if
   if pingmode = "refractory" AND counter = t_ping + t_refrac then
      set pingmode ← "inactive"
      set counter ← 0
   end if
   if pingmode = "inactive" then
      if any ping received? then
         set state ← active mode
         GENERAL MOTION BEHAVIOR(agentname)
         for i ← 1, no : ofpingsreceived do
            append list l ← heading of incoming ping
         end for
      end if
   end if
   if main_counter ≥ internal_timer_value then
      if is_illuminated? then
         set state ← active mode
      end if
      set main_counter ← 0
      set counter ← 0
      Create empty list, l
   end if
end procedure
procedure GENERAL MOTION BEHAVIOR(move agent)
   calculate h_mean of list, l
   set agent heading h_a ← h_mean
   move agent for distance β
end procedure
procedure AVOID MOTION BEHAVIOR(move agent, heading of detected agent)
   if agents in sensing range? then
      set agent heading h_a ← oppositetoagentinsensingrange"
      move agent for distance β
   end if
end procedure
```

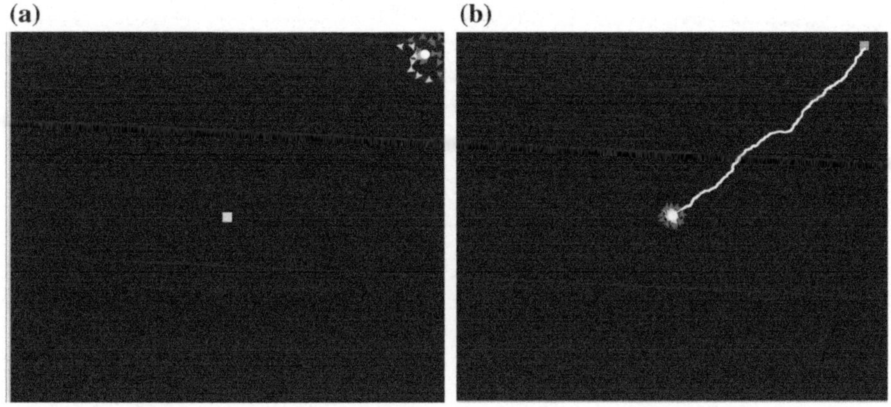

Fig. 4 The starting condition of a typical simulation run of the swarmFSTaxis is shown in Figure (**a**). Figure **b** shows a converged run. The green patch (occluded by the white circle in Figure (**a**) shows the starting point, the white circle shows the center of mass of the swarm and the yellow patch shows the predefined goal. The white trace in Figure (**b**) shows the trajectory of the center of mass of the swarm. The yellow agents are the illuminated agents and the blue ones are the shadowed agents

Table 1 Table showing all constants used in the modified swarmtaxis algorithm

Constants							
	t_{ping}	t_{refac}	s_r	β	$swarm\_size$	$avoid_{illum}$	$avoid_{shadow}$
Value	1	3	2.5	0.1	21	0.7	0.4
Units	ticks	ticks	p[a]	p[a]	–	p[a]	p[a]

[a]Unit p in this table represents distance unit in Netlogo

trajectory for the motion of the swarm as a whole in each run. The constants used for simulation are shown in Table 1.

In order to compare the swarmFSTaxis algorithm with the parent algorithm, we have performed 100 runs of each algorithm. To make the runs comparable, all runs were started from the same point, had the same swarm size, had the same parameters as shown in Table 1 and had the same goal. Also, the parameters used such as sensor ranges of illuminated and shadowed agents, range of directional communication and distance moved during motion behavior were kept the same for all 100 runs. Since the most intuitive comparison parameter is the time taken by each of these algorithms to converge to the goal, we have recorded and plotted this parameter in Fig. 5 for the swarmFSTaxis algorithm and the parent algorithm. Later, in Sect. 4, Fig. 5 is discussed in detail.

For testing the algorithms for their resilience towards agent-to-agent communication failure, we introduced a probability of failure, $0 < p_k < 100$, to each ping that an agent sent out. Before each agent sent out a ping, a roll of dice was done with probability, p_k to decide whether that ping should fail to be communicated. For each value of p_k, 100 runs were conducted and the number of iterations the

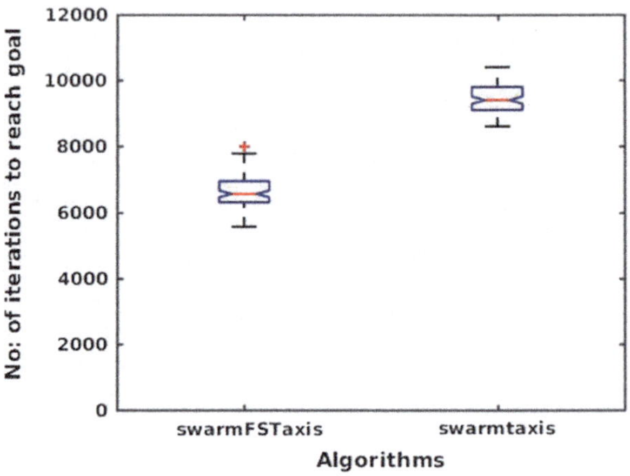

Fig. 5 The box plot shows the number of iterations each algorithm took to converge to the goal. The data from 100 runs of each algorithm is shown in the plot

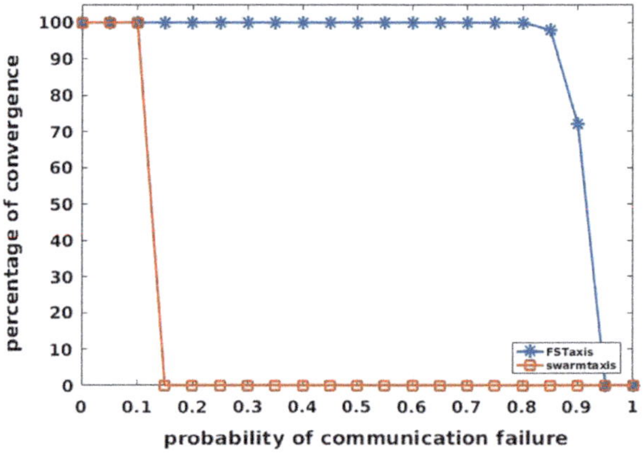

Fig. 6 The plot showing the percentage of runs that converged to the goal for all failure probabilities

algorithms takes to converge to the goal were recorded. Figure 6 shows the plot of the percentage of runs that converged to the goal for each failure probability. For both algorithms, 1,000,000 iterations were set as a limit beyond which a run was considered "non-converging".

4 Discussion and Conclusion

From Fig. 6, we can conclude that the resilience of the swarmFSTaxis algorithm has increased dramatically as compared to the parent algorithm. This phenomenal increase in resilience of the algorithm is due to the fact that the algorithm depends not on the count of pings for its state transitions but rather the direction of incoming pings. Even when as low as a single ping reaches an agent, the agent performs general motion behavior. Meanwhile in the case of the parent algorithm, if a ping from some of the agents fail to be transmitted or received, the algorithm executes sub-optimal transitions into the coherence state as it perceives that the swarm connectivity falls below the threshold α. The disadvantage of such an approach is that it makes the algorithm dependent on the success of communication of all the agents in the swarm. As shown in Eq. 1, assuming the probability of failure of each ping as p_k, there is a collective probability, $p_{coherence}$, that an agent counting the pings makes a transition into the "coherence" state. In contrast, the swarmFSTaxis algorithm fails in its state transitions only when all the agents in the swarm or almost all agents in the swarm fails to transmit or receive a ping. This probability is of the order of $(p_k)^{swarm\_size}$ and hence is very small for even very high values of p_k as seen from Fig. 6.

$$p_{coherence} = p_k^1 + p_k^2 + p_k^3 \cdots + p_k^{\alpha} \qquad (1)$$

In Fig. 5, we see that out of 100 runs of both algorithms, swarmFSTaxis algorithm is consistently faster than the parent algorithm. From Fig. 5, the mean of the number of iterations to convergence for the swarmFSTaxis and the parent algorithms are 6597 and 9540 respectively. Therefore, the swarmFSTaxis algorithm has become about 30% faster than the parent algorithm. As explained above, fewer transitions into the "coherence" state helps the swarmFSTaxis algorithm to be faster than the parent algorithm. In a later publication [10], the parent algorithm was improved and made more resilient to such failures, however, the state transitions were still based on a poll based count. In contrast to this approach, the swarmFSTaxis algorithm uses the direction of the incoming ping instead of a poll based count to keep the swarm coherent.

This paper demonstrates the improvements in resilience and speed that the communication mechanism inspired by fireflies and slime mold brought to the parent algorithm. In the future, the communication mechanism inspired by slime mold and fireflies may have the potential to replace poll based counts that are common in engineered systems. Further research in this direction can ensure that the full capacity of such a communication mechanism is utilized. Apart from using merely the direction of incoming pings, more statistical measures can also be developed according to the needs of the task to be accomplished.

Acknowledgements This work was supported by EU-H2020 Project no. 640967, subCULTron, funded by the European Unions Horizon 2020 research and innovation programmer under grant agreement No 640967.

References

1. Bjerknes, J.D., Winfield, A., Melhuish, C.: An analysis of emergent taxis in a wireless connected swarm of mobile robots. In: IEEE Swarm Intelligence Symposium. pp. 45–52. IEEE Press, Los Alamitos, CA (2007)
2. Hoff, N.R., Sagoff, A., Wood, R.J., Nagpal, R.: Two foraging algorithms for robot swarms using only local communication. In: 2010 IEEE International Conference on Robotics and Biomimetics (ROBIO), pp. 123–130. IEEE (2010)

3. Moeslinger, C., Schmickl, T., Crailsheim, K.: Emergent flocking with low-end swarm robots. In: Dorigo, M., Birattari, M., Di Caro, G., Doursat, R., Engelbrecht, A., Floreano, D., Gambardella, L., Gro, R., Sahin, E., Sayama, H., Sttzle, T. (eds.) Swarm Intelligence. Lecture Notes in Computer Science, pp. 424–431. Springer, Berlin, Heidelberg (2010). https://doi.org/10.1007/978-3-642-15461-4_40
4. Schmickl, T., Crailsheim, K.: Trophallaxis within a robotic swarm: bio-inspired communication among robots in a swarm. Auton. Robot. **25**(1–2), 171–188 (2008). https://doi.org/10.1007/s10514-007-9073-4
5. Siegert, F., Weijer, C.J.: Three-dimensional scroll waves organize Dictyostelium slugs. PNAS **89**(14), 6433–6437 (1992)
6. subCULTron: submarine cultures perform long-term robotic exploration of unconventional environmental niches (2015). http://www.subcultron.eu/
7. Thenius, R., Moser, D., Cherian Varughese, J., Kernbach, S., Kuksin, I., Kernbach, O., Kuksina, E., Miškovi, N., Bogdan, S., Petrovi, T., Babi, A., Boyer, F., Lebastard, V., Bazeille, S., William Ferrari, G., Donati, E., Pelliccia, R., Romano, D., Jansen Van Vuuren, G., Stefanini, C., Morgantin, M., Campo, A., Schmickl, T.: subCULTron—Cultural Development as a Tool in Underwater Robotics Consortium for coordination of research activities concerning the Venice lagoon system. In: Artificial Life and Intelligent Agents. Springer (2016) in print
8. Varughese, J.C., Thenius, R., Wotawa, F., Schmickl, T.: Fstaxis algorithm: bio-inspired emergent gradient taxis. In: Proceedings of the Fifteenth International Conference on the Synthesis and Simulation of Living Systems. MIT Press (2016)
9. Werger, B.B., Mataric, M.J.: Robotic “food” chains: externalization of state and program for minimal-agent foraging. In: Proceedings, From Animals to Animats 4, Fourth International Conference on Simulation of Adaptive Behavior (SAB-96), pp. 625–634. MIT Press (1996)
10. Winfield, A.F., Nembrini, J.: Emergent swarm morphology control of wireless networked mobile robots. In: Morphogenetic Engineering, pp. 239–271. Springer (2012)
11. Zahadat, P., Schmickl, T.: Division of labor in a swarm of autonomous underwater robots by improved partitioning social inhibition. Adapt. Behav. **24**(2), 87–101 (2016)

Teaching Complexity as Transdisciplinarity

Loren Demerath and E. Dante Suarez

Abstract This paper describes how a course in complexity studies can teach the value of a broad education, and the benefits of synthesizing knowledge otherwise acquired in disciplinary silos. For students focused on one field or vocation, understanding complexity may provide the necessary perspective that links their field to other valuable methodologies. Teaching complexity allows both students and faculty to connect disciplinary expertise to a wider range of knowledge on how things work, giving them a more consilient approach to solving real-world problems. The proposed course demonstrates transdisciplinarity across disciplines to identify self-organizing networks and the emergence of bounded systems. Agent-based modeling is used to show students how basic algorithms can create complex orders, and how lower-level orders can give rise to higher levels of order that have new, unpredictable properties. Students are given a grounding in both thermodynamics and information processing to understand how any kind of self-organizing system may evolve according the same generative principles, be it an ecosystem, a stream of consciousness, an industry, or a genre of art. As students see the prevalence of self-organization and emergence across disciplines, they can share with faculty the sense that we are in a special moment where a new, more unified way of seeing things is taking shape. Both authors of this paper have taught several courses on complexity, including one that was taught jointly with faculty from six different disciplines at five different universities. The authors remark on what worked, what did not, and what could be improved, as well as providing a set of recommendations and resources for faculty who may be interested in teaching on this important subject.

L. Demerath
Department of Sociology, Centenary College of Louisiana, Shreveport, LA, USA

E. D. Suarez (✉)
Department of Finance and Decision Sciences, School of Business, Trinity University, San Antonio, TX, USA
e-mail: esuarez@trinity.edu

T. Carmichael et al. (eds.), *Complex Adaptive Systems*, Understanding Complex Systems, https://doi.org/10.1007/978-3-030-20309-2_11

1 Background and Objectives

This essay argues that teaching complexity provides a unique opportunity for showing undergraduates the value of interdisciplinarity but also of transdisciplinarity, where disciplinary perspectives are applied in new ways to help answer questions traditionally pursued in other disciplines. Complexity can be defined as the study of the emergence and self-organization of networks of interacting agents. It has led to the development of concepts that together create a new perspective on such things as art, music, communication, governance, markets, language, consciousness, life, and the evolution of the universe. We urge professors and administrators to consider adopting complexity studies as a central topic of study for undergraduate students. Such students are often more willing to imagine new linkages between traditionally separated disciplines. Teaching complexity can also invigorate graduate level programs, introducing new methods and concepts to students, and potentially stimulating interdisciplinary collaborations among faculty. In the course of arguing for teaching complexity, we will share our experiences and the lessons we learned. We also offer resources to any interested readers that include our syllabus, lecture outlines, Power Point slides, class recordings, a concept map and glossary, class exercises, computer simulations and agent-based models upon request from the authors.

For both of us authors, and for many of our students, understanding complexity has provided a cross-disciplinary view of the world that is powerful. We will show some of that power here in the process of sketching the concepts that professors might want to teach as part of a complexity courses. As potential instructors take in those concepts, they can begin to imagine how their own disciplines and particular research interests might relate to this novel field. After all, the organic growth of meaning and complexity is everywhere in our world. Students will want to learn about how scientists and scholars are making sense of this tendency. Undergraduate programs offering a broad-based, or liberal arts approach to education are in a prime position to show it to them. Meanwhile, graduate programs stand to benefit from developing complexity curricula that show how methods and concepts developed in other fields complement their own research agendas. All disciplines study energy or information in some form, and in looking at how the focus of study self-organizes and evolves, advanced scholars may be surprised at how applicable complexity concepts are to their fields. Regardless of the level of instruction, though, studying complexity exposes us to new methods and perspectives, while also showing us how solving real-world problems requires more than a single-angled, dogmatic approach.

2 Emergence: Mother and Destroyer of Disciplinary Boundaries

Complexity is all around us. It permeates everything we do. Complexity is nature, complexity is us. Wikipedia currently cites Johnson's [40] book on emergence to provide a short definition of complexity as characterizing the "behaviour of a system or model whose components interact in multiple ways and follow local rules, meaning there is no reasonable higher instruction to define the various possible interactions." In looking for a precise definition, most everyone you meet who studies complexity may likely start by admitting that we currently do not have a clear definition for what complexity actually stands for. It's complex! As Waldrop's [73] *Complexity: The Emerging Science at the Edge of Order and Chaos* widely read popular science book admonishes, we are still "waiting for Carnot." This comment is made in reference to Nicolas Léonard Sadi Carnot, the nineteenth century French physicist and so-called father of thermodynamics, who brought a previously dispersed body of knowledge into a coherent whole. The field may indeed one day find its way to an all-encompassing and concise definition of what Complexity stands for, but in the meantime researchers in this growing field do not seem to be in a hurry to find it, at least partly because it is accepted to be an area of study in the making, and thus purposely intended to be open and welcoming to new methodological approaches and ways of thinking.

Often, the field is defined as what it is not: linear. Throughout the history of science, we have dealt with the universe's complexity by looking for simplicity, analytically distilling the common essence of things while hopefully retaining some descriptive realism [26]. The literature loosely refers to this school of thought as "linear" or "reductionist," meaning that it does not normally consider the structure or interactions between the parts involved [2, 20, 21, 57]. Think of the independence of events that one often finds (or not) in statistics. There should not be any experimental differences between throwing a die ten times or ten dice at once. Similarly, in fields such as Neoclassical Economics, the total demand for a product is simply obtained by adding up the demands of the individual consumers. Any time we are adding up agents, we are explicitly or implicitly assuming that there is no interaction between such agents, or that the interaction does not matter, perhaps because it cancels out in the aggregate. Another example is one of the most widely used statistical techniques throughout most fields of study, the *linear* regression, where causality is assumed to be direct and the result of adding up completely separable factors.

The simplifications of the past were generally adopted for purposes of tractability, and understandably so, since models grow exponentially more difficult to understand when we consider agent interaction [74]. Think of the way in which a standing ovation occurs in a concert [52]; it's not as if every single person attending makes a conscious, rational valuation of the quality of the performance with their eyes closed. No, we are influenced by others standing up, and there is pressure for standing when everyone around you has already stood up. These patterns of standing and sitting percolate through the system, mostly affected by local interactions, but with aggregate emergent effects. With the vast amount of computing power that is now available, researchers can model interacting agents and allow the computer to perform calculations that are impossible with traditional reductionist methods [34]. In contrast to a reductionist approach, the holistic, nonlinear view of complexity incorporates concepts that have been developed in several disciplines, and that allow for a richer, more realistic description of phenomena such as living organisms and evolving societies [48].

The nascent scientific consensus of complexity sees reality as a dance of interconnected elements, each of which cannot be correctly understood or described out of context [26]. The most common way of describing the difference between the linear and nonlinear paradigms is that in the former the aggregate is equal to the sum of its parts, while in the latter the aggregate is more than the sum of its parts [6]. According to the complexity paradigm, nonlinear systems cannot be reduced to their parts, nesting linear systems in the extreme case where interaction and adaptability of the parts is null [55].

The key word used by all in this field is *emergence*, which can be thought of as everything that cannot be reduced to the forming parts of the whole, as seen in isolation. Emergence is the reason why political scientists cannot describe an election in terms of the atoms involved. But what exactly is emergence? This is where the field has not coalesced to a proper definition. If the aggregate is more than the sum of its parts, where is that extra element coming from? One way of thinking about this is by recognizing that there is an element of 'design' that is lost when we try to define an aggregate as a simple collection of atoms or agents, without reference to how they are organized [1]. For example, in John Conway's game of life [4], there are three simple rules that describe the way in which "cells" are born and die. The rules are deceptively simple, giving way to incredibly complex configurations if the right conditions are set.

One example of emergence in the Game of Life is what is known as a "glider," a simple but moving self-sustaining structure that can transmit information across the lattice in which the game is set. The glider is not something that is explicitly encoded in the rules of the game, but rather a structure that emerges from it. But is it that we are simply not clever enough to see how the three rules of the game spell "glider"? Although Kauffman refers to self-organization as "order for free" [43], others point out that conditions of energy constrain it [31], just as conditions of information would constrain any person's perception of a glider. Does the perceiver know enough about what gliders look like to reduce the information of the lattice into that one image? New research by Lenartowicz et al. [47] integrate the work of Luhmann on systems [50] with Simondon on identity [67] to show how organizations show a kind of consciousness by reflexively acting upon themselves to craft their own evolving, compressed expressions of identity. Meanwhile, Latour et al. [46] start from the observation of early sociologist Gabriel Tarde [70] that "the whole is always *smaller* than its parts" (italics added), to show how the orders we construct can impede our understanding of their actual fluidity. The growing sense among researchers that everything is, actually, a network, parallels the rise of network analysis for good reason. How emergence occurs is an active area of research, and hopefully one of the universal mysteries that complexity studies will help explain.

In hearing about complexity for the first time, those who are not familiar with the field often ask about how it is related to Chaos Theory. It is related in that it is part of what could be broadly referred to as nonlinear science. More specifically, however, these fields can be thought of as opposites. In very simple terms, chaos can be reduced to "sensitivity to initial conditions," a concept popularized as the butterfly effect—the idea that small changes can have large effects in a system. In a layman's summary, the theory teaches us that simplicity can breed chaos. On the other hand, Complexity Theory hovers around the Second Law of Thermodynamics, which states that entropy always increases in an isolated system. In other words, one can tell time in the universe by the permanent increase in disorder it exhibits. We can thus think of complexity as the study of how order comes about in localized pockets.

In this last and general sense, thinking about complexity leads to questions about the nature of multidisciplinary thought. The implicit theme in complexity studies is that there exist commonalities in the way that this localized order happens in all aspects of the world we inhabit. The impulse towards self-organization appears to be universal and natural. Moreover, the idea of emergence from complexity teaches us why we have disciplines as separated areas of knowledge, how these disciplines are ultimately linked in a common reality, and how understanding that reality could be the next great frontier in scientific thought. As we learn about emergence, we will come to see our world as being composed of hierarchically-nested levels (see Simon's seminal [66] book, The Sciences of the Artificial, for an excellent treaty on the ubiquitous presence of hierarchical structures in nature and society).

Reality is multi-leveled, multi-faceted, and thus better understood in the hierarchical knowledge clusters we know as disciplines [68]. The laws, patterns and granularity of each level of description may not be applicable even in adjoining levels, and in this way we can understand our view of the universe starting at the quarks of quantum mechanics, followed by the atoms of traditional physics, the molecules of chemistry, the proteins of organic chemistry, the cells of biology, Darwin's fittest animals, the eusocial society of a beehive, and ending with the lack of a global government that can tackle global warming effectively. Emergence both helps us understand why there are distinct levels of description (disciplines) and how those levels interact with each other.

3 Interdisciplinarity and the Value of the Liberal Arts Today

Interdisciplinarity is in vogue these days. Universities are increasingly asking their faculties to develop interdisciplinary courses and research accordingly. The origins of this pedagogical research can be found early, as in the work of Binder [7], and later of Swora and Morrison [69], but there has been an increased focus on this type of approach to the scientific endeavor (see Nowacek [56]). Just like the definition of complexity, there seems to be a lack of complete consensus to what exactly constitutes interdisciplinary work, with new terms being developed to describe the way in which disciplines interact with each other, be it multi- or post-disciplinary. Rutting et al. [63] refer to interdisciplinarity in these terms: "*Interdisciplinarity has become a buzzword in scientific debates, and it has been identified by many research funding organizations in Europe and the United States as an important factor in future research. Although there is no single accepted definition of interdisciplinar-*

ity and the term is sometimes used interchangeably with multidisciplinarity and transdisciplinarity, it is important to distinguish and describe these three different manifestations of research."

Many have judged the American higher educational system to be the best in the world, although not free from areas in which it can be improved [38]. One of its most distinguishing characteristics is its liberal arts approach. Originally developed in the Greco-Roman world, it forced students to learn about grammar, rhetoric and logic before they could move on to other subjects. In other words, the view of the liberal arts approach is that there is a layer of knowledge behind all disciplines, that it is common across fields, and that studying it can help anyone trying to understand the way our world works. Although originally adopted in the United States in modern times, the approach is now being emulated around the world, and adopted at least partially in more traditional American universities. The rationale for teaching liberal arts is that students will have greater rates of success if they possess diverse skills [65], with an education that involves taking courses in disjointed disciplines to develop the critical thinking skills of a well-rounded graduate. That belief is based partly on the unpredictability of how students will have to solve problems in their future.

For the purposes of this essay, we refer broadly to a liberal arts education as simply one in which students learn from different disciplines, regardless of their area of study, stressing the fact that there are multiple ways of studying our world, all being potentially just as valid. We distinguish this general approach to a broad education as liberal arts, without capitalization of the concept, and keep the capitalized version of Liberal Arts to refer to mostly American colleges that adopt this view wholeheartedly. Thus, the liberal arts approach can happen both in a Liberal Arts university per se, or in a more traditional university or program that adopts this type of pedagogy. In institutions of higher learning in the U.S. and around the globe there is a broad range of variation between the holistic approach of the liberal arts and a education based on a single discipline or paradigm. To bound the range of this educational dimension, we can place St. John's College, in Santa Fe, NM, in one extreme, and the Chicago School of Economics on the other. At St. John's there is only one major, Liberal Arts, where students learn about the traditional disciplines by reading the original works of Newton and Marx. The University of Chicago School of Economics, on the other hand, concentrates its instruction on the highly mathematized paradigm of Neoclassical Economics, with: "A deep commitment to rigorous scholarship and open academic debate, an uncompromising belief in the usefulness and insight of neoclassical price theory, and a normative position that favors and promotes economic liberalism and free markets." [44].

In any given domain of experience in reality, everything unfolds in unpredictable ways, and our world has become increasingly complicated and interconnected. Laboratory manuals or engineering guides for how to deal with problems in the lab or with a machine are insufficient for dealing with the intricacies of real life. Consider for example a small firm that wishes to release an app on the market. Questions of computer design, marketing, psychology, and visual design all play a part in the finished product and all of these interconnected structures need to be working in tandem to have a successful release. Students need to be creative, and the more diverse approaches to knowledge they are exposed to, the more creative they are likely to be.

Part of teaching liberal arts, though, has to be teaching the value of learning and deploying that diversity. The value of diverse approaches has been demonstrated through outward application and inward fit. The former case is characterized by an etic, or idiographic approach of case study where educators use a broad range of approaches to understand a particular place or moment in time. For example, Cortese [15] describes how Unity College in Maine focuses its curriculum on understanding nearby Lake Winnecook. A personalized approach has been to allow students to fit the curriculum around their own interests. It is now relatively commonplace for schools to allow students to craft their own interdisciplinary majors [35, 42]. This is particularly true for Liberal Arts schools, such as ours, which have sought to maximize the benefits of interdisciplinary thought, with the hope of finding ways to promote collaboration across different curricula and encourage faculty to find productive coalitions.

Graduate programs at research oriented institutions may stand to benefit the most from teaching complexity. Understanding the common themes of complex systems connects researchers in disparate fields, and through those connections gains are turned into novel methods, concepts, and findings. Further, with no single theory unifying the phenomenon of complex systems, this is fertile terrain for projects with theoretical aspirations. Progress on that front should yield a common vocabulary that would accelerate research programs and stimulate new fields. In this way, learning about complexity improves our interdisciplinary connectivity, and researchers that are more diversely connected are likely to be more innovative and productive.

4 Complexity as a Bridge of and to Interdisciplinarity

The value of a liberal arts education has been attacked in recent years, particularly in the context of college debt and the employability of a college graduate (see [3, 25, 49, 65]). Michael Lind, as an example of those defending the approach states, *"Liberal education in some form will survive as long as societies need not only leaders but also ordinary citizens who know how to read, write, and reason."* But often lacking in descriptions of the benefits of a liberal arts education is how it actually achieves these objectives (for a good counterexample of this, see Reybold and Halx [62]). Moore and Robert [53] note how students are better equipped to face challenges with diverse skills and knowledge. Seifert et al. [65], find positive effects of a liberal arts education on openness to diversity and challenge. In this essay we argue that complexity studies allow us to go beyond obtaining knowledge breadth to understand how it coheres. For this type of undergraduates, a liberal arts education can lack the common glue between the different courses that a typical student takes, as well as a way of connecting the student's major and the rest of their education. We propose that complexity can serve as a way to bridge these gaps.

For more advanced students and faculty, however, complexity's benefits to research and understanding can be just as significant. At the Santa Fe Institute, arguably the birthplace and currently the mecca of Complexity studies, people speak of their approach to doing science as post-disciplinary. They believe that in order to understand the underlying phenomena describing the many levels of reality of a particular complex system, a team must be formed with scholars working together from multiple disciplines. A prominent researcher of the Santa Fe Institute is Scott Page, who teaches at Michigan University's Center for the Study of Complex Systems. He speaks of an approach to science that deals with the multiple levels of reality through "many models." In other words, the idea that he and others propose is that if we are to understand the causes, effects and policy challenges that a phenomenon such as income inequality poses, we must be ready to view it from multiple perspectives (see, for examples of this multilevel approach to modeling, Brown et al. [8, 9]). Is income inequality more than simply an economics issue? Is it both a micro and macroeconomics phenomenon? Does sociology have anything to say about it? Can an anthropologist provide a multi-ethnic and historical perspective of the issue? Should a political scientist care about it? Is the process of increasing inequality actually parallel to the Second Law of Thermodynamics and its ever increasing entropy? We believe that the answers to all of these questions is definitively affirmative.

In studying any significant social issue, various points of view will be rooted in particular disciplines, and that calls for a common language in which the different views can be brought together. In the context of doing this kind of interdisciplinary research, we must have a more open approach to the value of scientific inquiry and, in particular, to the value of modeling. Epstein [27], explains in his influential article "Why Model" that the purpose of modeling is not just to predict the future. He eloquently points out that any time we attempt to understand a complex phenomenon and explain it, we are implicitly modeling it, "*...but typically, it is an* ***implicit*** *model in which the assumptions are hidden, their internal consistency is untested, their logical consequences are unknown, and their relation to data is unknown. But, when you close your eyes and imagine an epidemic spreading, or any other social dynamic, you are running* **some** *model or other. It is just an implicit model that you haven't written down*." When viewed through this perspective, modeling can be understood as an integral component of rational thought. Epstein argues that among other benefits, modeling allows us to understand the problem better, to figure out exactly what it is that we should be measuring, to discover new questions, to bound the possible outcomes, to challenge the robustness of prevailing theory through perturbations, to train practitioners, to guide the policy dialogue and educate the general public.

Of all the types of modeling that could be done, the most common modeling techniques of complexity are particularly well-suited for this type of inter and transdisciplinary conversations. Complexity allows us to see the commonalities across disciplines that together reveal essential aspects of how nature works, in all kinds of systems with emergent properties; physical, biological, ecological, psychological, cultural, economic, political, even aesthetic (see Geels et al. [28], for an example of Society and Technology co-evolving). Complexity studies reveal that there are common structures and processes underlying organic, naturally arising, self-regulating systems and phenomena, and that they can be studied. The process of doing so helps students and faculty better understand general principles and allows them to see a discipline's perspective in a more naturalistic way, with greater awareness of how it is evolving and how it will continue to evolve.

For example, when someone is studying biology and evolution, she may not realize that the principles at play are much broader than just life on this earth. The recent explosion of work in the area has shown how the laws of evolution are simply one manifestation of more general principles in energy and information processing (see one of many examples of how multilevel selection theory affect the evolution of cooperation, [17]). Complexity therefore generally points to that which is lost when one is too focused on a particular question. Eventually, interdisciplinarity may become its own discipline, but until then, complexity can serve the purpose of identifying common themes across disciplinary and methodological approaches to science. In terms of higher education, one model for broad-based education we currently have is the Liberal Arts college. But even this approach can be revamped, not as the teaching of many different disciplines separately, but in focusing on how these different disciplinary points of view fit in the larger picture. For example, Trinity University, a Liberal Arts university with professional programs where Suarez teaches, recently revamped the common curriculum of all its students to include an "interdisciplinary cluster," where students take three related courses from different academic departments and are forced to see the interconnections between them.

Teaching complexity should be appealing to many academics. In giving faculty a set of common concepts and shared terminology, teaching a course in complexity such as the one we are proposing would enhance the professors' capacity to connect to others outside their fields and subfields [51]. Furthermore, teaching in a field that is not fully cemented forces the instructor into a position of humility, where she teaches the students how to learn and incorporate other points of view, more akin to a true scientific dialogue than to a simple instruction where students absorb knowledge from faculty without having much opportunity to question it. As all of us professionals in the field know well, we are encouraged as university professors to continue to be active scholars throughout our careers, as, we are told, research allows us to stay current in our fields—something which is of particular value in the classroom. Active research keeps us fresh; it keeps us honest about what we teach.

Humility is central to complexity studies, as the pervasive idea of emergence reinforces the fact that a model may only describe one level of reality. In the same way that when we implement and apply any model we must be aware of its assumptions and limitations, work in complexity often makes explicit the levels in which it is applicable. Emergence helps us understand how many principles and methods that are pervasive in any one discipline may only apply there. What is so important in one realm, is irrelevant in another. They key variable in one model may be little more than an irrelevant blip in another. Complexity also reminds us that the disciplinary boundaries are fluid; as active scholars we must always be aware of methodological and theoretical innovations that may change the relevance and applicability of our models and theories across fields of study.

As we develop models of emergence, we better understand how disciplines are linked to each other. By shining a light on emergence itself, and explicitly modeling it, different realities are depicted and explained simultaneously with a single, overarching framework. In the blossoming field of Agent-Based Modeling researchers create simulations that discover self-organizing orders with emergent properties that reliably mimick real world patterns. These type of systemic and multi-leveled models can help us achieve consilience between disciplines, particularly adjacent ones, like economics, political science, sociology, and psychology (see, for example, Rensicow and Page [61], for an example of complexity applied to health care). In pedagogical terms, we want our students to be able to look at problems without any preconceived notions of what disciplinary methodologies to apply. We want our students to be able to tackle a real-world problem holistically. To develop a discipline, in a sense, for each problem. In contrast to this approach, clumping scholars who think alike into an isolated discipline leads to confirmation biases and a self-reinforcing building of dogmatic orthodoxies that limit what one can see and understand.

Furthermore, in the era of complexity, the edges between disciplinary nodes are increasingly becoming active areas of research, such as econophysics, behavioral finance, and neural anthropology. Biological evolutionary thinking now permeates Computer Science, Economics and Engineering. A side effect of these interdisciplinary studies is perhaps its revelation that the current boundaries between disciplines are not necessarily where they should be. For example, economics is a discipline that became highly mathematized in the 20th century, leading the field away from its more interdisciplinary incarnation of what was then and now known as Political Economy, where economics fully embraces its social and humanistic origins and realities, addressing issues such as ethics and culture. In contrast, an attempt to connect the macroeconomics level to its microeconomic foundations analytically, lead the field that came to be known as Neoclassical Economics to make very strong and unrealistic assumptions regarding the rationality, independence, and exogeneity of its idealized consumers and firms. Once again, these assumptions were adopted out of necessity, since this simplified Newtonian approach was the only one that at the time could deal with the analytical complications of building a mathematical model of the economy. This process led the discipline to relative isolation, particularly creating daylight between it and sociology and psychology. And such issues are not limited to the field of economics. A direct historical analogy could be made with the field of Biology, which took its time to accept ideas of multileveled selection theory [16, 76]. The field of biology was thus affected by the similar technical and conceptual deficits as economics in the twentieth century. Through complexity, students in these fields can now see and participate in science in the making, where in the state-of-the-art of scholarly activity these disciplines are meeting each other to create new knowledge (for active examples of interdisciplinary science creating consilience, see Gersick [29, 77]).

In the last decades, areas of economics have begun the process of re-connecting themselves to experimental, social and psychological methodologies, getting the field closer to consilience, that is, to a seamless transition between its body of knowledge and that of adjacent disciplines [75]. Some of economics most famous Nobel prizes winners are mathematicians (John Nash, 1994 awardee), interdisciplinary firebrands (Herbert Simon, 1978 awardee), or psychologists (Daniel Kahneman, 2002 awardee, along with Amos Tversky). According to the press release of the Royal Swedish Academy of Sciences, the most recent Nobel prize recipient, Richard Thaler, won the award for "Integrating Economics and Psychology." While this new, more fluid take on economics is not quite mainstream, we hope that more and more undergraduate courses will teach this broader view of economics, and not relegate these essential connections to a late chapter in the text book that the actual course never reaches.

In summary, we posit that students who learn the generalizability of their knowledge will be more likely to engage issues that lie in cross-disciplinary boundaries and demand diverse skills and perspectives. Real-world phenomena are often messier than those studied from contained academic views [5]. For example, a biology major may later work to solve environmental problems. If she understands the holistic underpinnings of the methodologies, principles, and laws learned in her discipline, she will be better able to apply them to the human realm of decision-making, such as when they decide to best manage a fishery. Survival of the fittest may allow her to understand that inefficient political regimes are less robust; or that an unfair distribution of resources can spell doom for a fishery.

5 Transdisciplinarity and Other Benefits to Teaching Complexity

Wikipedia refers to *transdisciplinarity* as a "*research strategy that crosses many disciplinary boundaries to create a holistic approach. It applies to research efforts focused on problems that cross the boundaries of two or more disciplines, such as research on effective information systems for biomedical research (see bioinformatics), and can refer to concepts or methods that were originally developed by one discipline, but are now used by several others, such as ethnography, a field research method originally developed in anthropology but now widely used by other disciplines*" [54]. This post disciplinary approach is fully embraced by complexity. As stated previously, the reality that of complexity has not been fully defined is probably due to the fact that many disciplines use its precepts in slightly different ways [39]. Another one of the reasons is that this is indeed a new field, and all the doors have not yet been opened.

In teaching complexity in such a transdisciplinary setting, we are showing our students the birth of a new paradigm, allowing them to experience science in the making. Also, because it would be hard to find someone with traditional disciplinary knowledge across many fields, a course teaching such concepts will either be explicitly co-taught by faculty from different academic departments, or at least have faculty and practitioners from other fields come to give guest talks. In this way, more faculty members get to participate in that genesis and can get inspired to find new research avenues, potentially bringing together academics from disciplines that do not normally interact, and creating productive collisions with these encounters. As mentioned previously, there are often only a few colleagues in any given discipline at small universities, such as Liberal Arts colleges like the ones the authors teach at, so to be productive researchers, professors have no choice but to collaborate with those in different disciplines. Such an environment provides students with unique opportunities to see faculty interacting across fields. Students also get to see the faculty themselves learning, allowing for a first hand introduction to the joys of intellectual curiosity.

Furthermore, studying complexity in a transdisciplinary setting opens a door to emerging methodologies such as network analysis, advanced statistical methods, simulations, big data, cellular automata, multi-agent systems, and multi-leveled modeling. These traits represent marketable tools for our students. Any graduate in the 21st century will likely benefit from knowledge of computer modeling, critical thinking skills, even intercultural capabilities—all of which are central to complexity studies. Students and faculty alike can identify payoffs of using such a perspective for a proper balance and interplay between closure and openness, between simplicity and realism in a chosen model. In working with a group of researchers, participating students and faculty also learn about the benefits of working in groups, particularly groups with diverse backgrounds and points of view [19, 58].

6 Establishing Common Ground in Course Preparation

A course on complexity co-taught by the authors of this article exemplifies how transdisciplinarity can work pedagogically. Preparation of the course was funded by a grant from the Associated Colleges of the South awarded to co-principal investigators Loren Demerath, one of the authors of this article, and Mark Goadrich, a computer science professor at Hendrix College. The grant included the participation of Dante Suarez, the other author of this article; Steve Desjardins, chemistry professor at Washington & Lee; Scott Davis, professor of philosophy and religious studies at the University of Richmond; and Jeanne Hamming, English professor, also at Centenary College of Louisiana with Demerath. During the course's preparations and its pilot instruction, the group met online, via email and bi-weekly video conference calls, as well as in-person during two different "summits."

Early in the course's preparation, we faculty held a summit where we all met at the University of Richmond. We took turns presenting on our choice of aspects of complexity which we found particularly valuable. In the course of those presentations, we could see that we each thought and talked about complexity in different ways, and yet there were commonalities to what we were each talking about. The highlight of our discussions on that first day was probably characterizing what those commonalities were. Indeed, that was the work of much of our meetings, at both summits and online. We would refine our concepts and terminology such that they would work for each of us; working, as we were, in different places, but somehow on the same thing. We may not have agreed that what we were feeling was an elephant, but we could tell we were feeling the same thing.

Finding each other as kindred spirits was one of the joys of our project; not that it took much work. Indeed, it is a testament to the timely importance of the complexity perspective that colleagues were so easy to find. Demerath had a geologist and English professor to choose from on his own very small campus. The latter, Jeanne Hamming, had studied how literatures in both ecology and science fiction demonstrated the logic of complexity, in the sensitivity of ecosystems and how small changes can have large, unpredictable effects. Through this experience, we were able to see each our subjects from different perspectives, to explore our disciplines' connections to each other, and to see our pedagogical styles contrasted and complemented. At first, Davis was not especially familiar with complexity studies, but he was ready to learn, and in the course of our project became increasingly enthusiastic for the way it helped him make a point he had been seeking to solidify for some time, that being the necessary imperfection of communication and the importance of a pragmatic approach. In one of the many very fruitful interactions, for example, Goadrich developed an agent-based model simulation of a simple economy, which Suarez used to talk about income inequality as a result of entropy. Hamming and Davis always kept our discussions grounded to proper epistemological and historical perspectives, and Desjardins kept us real with phrases such as "no one has ever seen the Second Law of Thermodynamics being broken."

Another joy of preparing and teaching the course was the challenge of learning each other's perspectives, and in teaching our own to colleagues with different backgrounds. Because a transdisciplinary approach can involve borrowing from each other's disciplines to address common issues, these meetings were not only important, but intellectually intense and fun, as we were each learning from one another. We maintained that feeling once we got to the course, allowing students to see the hedonic rewards of the intellectual work inherent in bridging disciplines. Indeed, it is this very same joy to be found in shared scientific research across different disciplines that a complexity course would seek to impart. Demerath would engage with each "guest" instructor, a bit like the over-eager graduate student so many of us once were. As we prepared for teaching the course, we had to learn how each other thought about certain topics, terminology, and the techniques used to demonstrate their utility. Sensitivity to initial conditions, for example, was illustrated through both poetry and science fiction narratives, by the English professor, by computer models of cellular automata by the computer scientist, in examples of countries and their underdevelopment by the economist, and by thermodynamic equations by the chemist. That meant that at some point in the course, students and faculty would all engage with Emily Dickinson, computer models, differences in historical economic development, and differential equations.

Certain recurring concepts kept us centered on that conviction. In each of our fields we could identify self-organizing systems with emergent properties. We could see the omnipresence of the second law of thermodynamics in energy dispersal, information decay, death, memory loss, or the decline of value with the loss of novelty. Also easy to spot was the tendency to process information by sorting, driven by laws of attraction, homophily, and the greater efficiency of interacting with similar others. As we shared our perceptions of these kinds of phenomena in our different fields, we developed a rough consensus on the meaning of complexity and its key concepts. In the following section we will describe how those concepts were developed and linked together in the sequence of the course.

However, part of the power of a live, interactive education is the spontaneous, unpredictable dialogue over the material between students and faculty. Helping motivate that dialogue is often the enthusiasm and detailed background knowledge of the instructor. Because both knowledge and enthusiasm will vary for each instructor, such courses work best when they are designed by the instructor herself. We offer the following account of our course not as a prescription for how to sequence the introduction of concepts or link them together, but as a way of visualizing one way of doing so.

7 An Undergraduate Course on Complexity and Transdisciplinarity

The plan of the course was to introduce students to the general idea that complexity exists across disciplines, then visit the fields typically examining those instances one by one in the course of the semester. At the semester's end, students would present their own papers on a particular instance of complexity, often in the disciplines of their own majors. As we proceeded through the course to that end, the different faculty would "visit" class through a video conference call, with their faces projected onto the screen at the front of the class.

Our first off-campus expert to visit us on screen was Steve Desjardins, a physical chemist who guided us through two weeks of complexity related natural science. Our course led with the big picture: complexity is about more than just the evolution of life, it is about the evolution of anything that evolves, and that includes a lot, both before the first genes and after the first memes. To have students grasp the span of complexity, we started with selections from David Christian's *Maps of Time* [13], which presents a sweeping overview of history. A leading work in the new field of "Big History," it begins just before the big bang, getting to humanity about a third of the way through the book, and ending with speculations on the next 100, 100,000, and 100 million years. Christian provides accessible context for the premise of many complexity scientists that surveying history over that time allows us to see an evolution away from entropy (to paraphrase Stuart Kaufmann in Waldrop [72]). Selections from the book summarize common patterns across history, and even phenomena, such as the way different kinds of agents of similar energy or information gravitate towards each other. Such is the case when particles of mass in outer space join to create stars, then higher elements, then planets and solar systems. Likewise, human beings are similarly attracted to the shared information of others, moving from rural to urban spaces, and gravitating to those who share their level of resources and culture. While attraction to shared information is manifested in the isomorphic forces of physical gravity and social homophily, energy and information dispersal is likewise seen in both phenomena.

The second law of thermodynamics, which states that free energy always disperses towards entropy until it reaches equilibrium, can be seen throughout geological and human history as driving change. So, too, can forces of attraction. Some forces are physical, such as gravitational, electromagnetic, and the strong and weak nuclear forces, while others appear to be based in meaning and are social. Indeed, Christian pauses in his survey of "big history," and cites the need for a theory of social gravity which could make sense of the way cities grow much like stars. Formalizing such a theory has been the goal of Demerath's recent work, and that was one reason he was excited about assisting the physical chemist colleague in the next section of the course. Demerath was hoping for Desjardins' approval as a physical chemist on his interpretation of certain aspects of thermodynamics and energy correctly.

The students too, were ready to meet Desjardins, and approach him from the same "transdisciplinary" perspective; to learn from his natural science expertise on such topics as chaos and dynamical systems, in order to eventually apply them to phenomena typically studied outside the natural sciences. Their paper topics were on their minds already. They knew they would have to apply these upcoming concepts—somehow—to their diverse topics. Like their lead professor, they had some skin in the game.

With his thorough introduction, Desjardins gave us tools to play with complexity. Such is the lead physics has taken in complexity studies. With a few simple rules leading to complexity, those rules start in physics, the "simplest of the natural sciences." From our concept list shown in section VIII, one can imagine how intimidating those covered by Desjardins might have appeared to some of the students not in the natural sciences. The concepts include energy, gradients, thermodynamics, phase states and transitions, information theory, computation, cellular automata, strange attractors, differential equations, chaos, points of criticality. There were some equations described, the concepts are abstract, and the readings assigned for those classes were highly technical. But while it was all confusing for some, others found it easy, even remedial, as they were already familiar with many of the concepts and terminology. As those students admitted that to the class, though, it seemed to encourage many of the non-science students. Many of them then reread the notes, determined to understand what their classmates found to be so straightforward.

Thereafter, the class was easier for the non-science students, and at times it was the science students who struggled. Those less experienced in psychology—the course's next section—or in any of those that would follow—sociology, economics, linguistics, literature, or philosophy occasionally found the going difficult. But students and faculty alike became accustomed to this aspect of the course: alternating between the role of expert and novice, encountering the familiar and new, respectively. But each time, complexity was a common ground, on hand orienting us to new material with familiar concepts and agendas, and on the other hand pushing us on old material with new questions. In other words, familiar knowledge would be put to new use, and new knowledge would be used on familiar topics.

By the time the class finished its sections on physics, computer models, biological evolution, and ecosystems, students knew that the study of complexity is about how agents of energy and information—be they atoms, organisms, or species—self-organize into systems with emergent properties. We had also established that those systems are ever-changing, since all energy must disperse, as described by second law of thermodynamics, including all order and information. But we had also grounded ourselves in the forgiving, comforting awareness that the change and fluidity created by death and decay is the stuff of life.

During this natural science section of the course, Demerath engaged Desjardins on how life and death are related not only in the literal sense, but at the more general level of attraction on the one hand, and energy dispersal on the other. Demerath suggested that change makes for difference and novelty, and those are the essential ingredients for both life and meaningfulness; difference provides "gradients" of free energy and meaningful information. Desjardins acknowledged that something like that could be at work, and that was enough for Demerath. Demerath shared with the class how in a private conversation he had asked Desjardins what the explanation is for why the chemical elements evolved as they did, in the order they did, and he said, there really isn't one. Demerath shared his excitement with the class at that answer: here was a big question unanswered that complexity studies could answer! What was previously an impassable obstacle became another avenue for scientific thought—a daunting but doable task. Such is the power of complexity.

And so, with Desjardins' general permission, the class moved away from the question of evolution and life with the understanding that a broad class of systems might display the same characteristics of "living" at points of criticality: balanced between life and death, order and chaos, depending on some degree of change for energy and information processing. All such systems self-organize, grow, and evolve, and they all beget new, unpredictable orders beyond themselves. Those new orders emerge in a hierarchical fashion to constrain the lower orders, but because their properties are emergent and unpredictable, the higher orders cannot be reduced to the lower ones, nor the agents that constitute them. Such as our thoughts, that constrain our actions; or our society, that constrains our thoughts. Such were the topics of our next few weeks.

Demerath then stepped out of his role as the over-eager graduate student, and again took the part of class instructor. For the past two weeks he had been supporting/pestering the chemistry professor with questions and comments of affirmation, confusion, reiteration, or acclamation. Now they said goodbye to Desjardins (who said he was surprised with how well the class went during his time), and Demerath guided the class into its sections on psychology and sociology.

On the heels of discussing evolution, we go to the head, to address the first constraint on human behavior to emerge: conscious thought. Students were first asked to consider the purpose of conscious thought, and how consciousness developed and evolved. A variety of ideas encouraged the class to think of various ways in which the mind might gain efficiency in processing information. Students were exposed to the research summarized by Daniel Kahneman [41] suggesting that conscious thought is a kind of focused "slow" thinking to which we give full attention, but which is distinct from "fast" thinking that goes on in the background of our minds and of which we're largely unaware. Students were asked to imagine what might have caused us to be able to concentrate our thinking on more important features of the environment, be it physical or social.

One of the most mysterious moments of emergence of a higher level system is our ability to sense our own existence. But consciousness has recently been made less mysterious by using a complexity perspective. Christof Koch is a neuroscientist who has summarized key insights into how consciousness works [45]. One of them is how the neurons most closely identified with consciousness spend most of their activity turning down sensory input, not passing it on. Those neurons are filtering information, rejecting most of it, but accepting some to pass on to the neural networks that we know are involved in conscious thought.

At this point, students were asked to consider what the benefits of excluding information might be, how such sorting is actually information processing, as well as the ways in which sorting can happen. Demerath then presented John Hollands' image of "gated urns" [36, 37], where objects floating in space are attracted to the entrance of some urn-like container, but aren't admitted unless they contain enough shared information as the objects inside. Such a simple mechanism eventually yields a more homogeneous environment on the inside of the urn. That improves the efficiency of communication of the agents, who no longer have to "ask" if certain information is shared.

To create those environments, then, we look for shared information. That includes our thoughts. Students are reminded that our brains are thermodynamic systems, and research has shown that our brain, like the brains of all organisms, are entropy reduction systems that follow the laws of thermodynamics [30]. This means brains are systems that seek to reduce energy dispersal by ordering thoughts and understandings, and thinking in orderly ways, going from thought to thought based on similarity, but exploiting differences in novelty as energy gradients and information. We are motivated to process information efficiently, be it information about our environment, or about our information itself. So, changes in the way we think about one thing can affect changes in how we think about something else. To process this information efficiently we focus on our thinking. Conscious thought, then, is really a self-organizing information processing system, that, like any other, exploits gradients of difference between similar agents of information to reduce those differences, categorize the agents, thereby assimilating the information, and reducing entropy. And like other such systems, consciousness exists at a point of criticality between chaos and order. This is why we are most likely to be conscious when we are taking in information. Like a water wheel, where the difference in water levels turns the wheel, so does the difference between what we see and what we expected turn our wheels of conscious information processing.

We expose students to such recent conceptualizations of consciousness (e.g., Tononi's Integrated Information Theory [71]) that see consciousness as a state of information being held in existence by a certain level of information processing. That state where grouped elements or representations make a difference to each other through feedback loops, including models of the environment and the environment, or, potentially, representations of the environment with each other within that model. Students are asked to grapple with the notion of "intrinsic information," where information is valued not for its ability to represent the environment external to us, but for its ability to order our internal cognitive environment. We show students research that approaches consciousness from a thermodynamic and information theory perspective, such as that of Collell and Fauquet [14]. It can take more energy to jump between more disparate things, including in our sequence of thoughts.

Students were encouraged to be critical of this view, and to ask questions like: what would the thinker's motivation be to think this way? Why not think more boldly, with greater tolerance for inconsistency, and less desire for similarity? Why bother even thinking at all? Why don't we rest, save energy, and if we have everything we need, stop thinking altogether? It would be functional, after all. Why should we think like information processors all the time? Is it just because it's functional? Aren't we told to be wary of functionalist theorizing? After all, it is easy to think up benefits to some situation, but that does not mean those benefits are why it evolved to be that way. As any sociologist can tell you, such "functions" are often just rhetorical devices to justify keeping a status quo that is suboptimal for most, but suits the elites just fine.

Demerath affirmed that suspicion and encouraged students to consider the basic rules guiding agents of energy and information, then, to describe what thinking would look like if thoughts were such agents, attracted to others with shared information, but also seeking some novelty, providing exploitable gradients of difference for information and energy. It would look something like what we do, and we did not have to explain how it was functional. Indeed, as students were asked to talk about information they know and find entertaining to think about, the a-functionality of those domains is striking: news stories we follow, shows we "stream," sports we watch, celebrities we find interesting, none of it is important information; while it might function as fodder for conversation, in itself, it is not important we know it. What we enjoy is to talk about it, to put it in order, and reduce the chaos of the world we see Demerath [22]. That is information processing for its own sake. And that is our thinking as well. Sifting through the inputs we gather from the environment, our memories of what happened, plans for what could, we make sense of them, order them, link them to something related, or, if at a loss, look around something new. And when things fit especially well, we gain significant information integrity and "predictive power," and this feeling of fit is aesthetically satisfying [24, 23]. Indeed, aesthetics have been thoroughly documented as a response to gains in information fluency [60].

At that point in the course it was hard for Demerath to resist playing a drum roll and cymbal crash and then announce, "Ladies and gentlemen, there is our link between objective conditions to subjective experience: the "good" aesthetic sensation of order on the one hand, and the "bad" sensation of disorder on the other!" There may not have been ooh's and aah's in response, but this did prove a useful reference point moving forward into the worlds of culture and valuation, particularly at the very end of the course, when we discussed religion and ethics with philosopher Scott Davis.

Before jumping into those social worlds, though, we needed to understand humanity's basic means of being social in language. For that purpose, Davis made his first appearance in class. The highlight of that set of classes is unmistakable. It was Davis' clear and compelling instruction of Donald Davidson's position that there is no single language that we all know, as we each have different understandings in our individual "idiolects" of what seem to be the same words, and even those understandings are changing, moment to moment. As such, not only do we never completely understand each other, one could say that we never completely understand ourselves. But while there is no such thing as complete shared information, there is enough there to process new information, a gradient of difference that energizes the interaction. Thus, we find repeated moments of pragmatic agreement. Agreement that is close enough to work with. Davis referred students to the long history of pragmatism in philosophy that is complemented by Davidson's work. Demerath would now also suggest noting the recent move by philosophers such as David Danks [18] to understand meaning and language as better represented through networks of graphical models rather than discrete rules. Doing so allows us to understand meaning as a function of strength of connections, or, in Danks' terms, "relevance." In making this shift, it would appear we can account for variations in language use that we cannot do otherwise. Future iterations of the course could well use selections from his book which accessibly describes the principles behind network analysis without using mathematics.

And network analysis is indeed a critical tool of complexity studies. The idea that the whole is bigger than the sum of the parts rests on an understanding of emergence as a function of the relationships between agents, not the properties of the agents themselves. Understanding the power of both informal cultural norms and more formal social institutions to regulate and coordinate relationships is essential to attaining a sociological perspective. Demerath prodded students to consider various forms of the most basic questions in applied sociology: Should you be social or not? Should you make that connection or not? Should your group admit or exclude that person? Should you diversify or self-segregate? When is it better to focus on what you know, and when is it better to explore what you don't? As sociology reaches a point of maturity in its development of its own network analysis techniques, in conjunction with adopting conceptualizations developed in complexity studies, it will likely be able to answer those questions with more precision and confidence than it now can [64]. That should lead to more effective treatment of social problems, and a better understanding of social puzzles. For example, it is difficult to grasp why it is o.k. for a subordinate culture to appropriate from a dominant one, but not vice versa, and even what makes a culture subordinate, without understanding the power of position within a network. Moreover, the inimitability of position in time and space for any kind of agent, including cultures, means that what works for the dominant culture may not work for any other. And, when the dominant agent cannot recognize the misfit on its own, a pragmatic agreement may be bypassed, and the result is a disconnection and a waste of energy and information, not to mention the sense of social justice at the heart of a healthy community.

The implications of power for determining not only economic value, but a hegemony of values and arrangements that support it was a chief concern of economist

Dante Suarez, when he then took over the course. "Technological lock-in" describes a situation where a dominant agent can arrange relationships in its own interests, and a network's growth can be stifled, a technology's evolution constrained, etc. Were that not the case, we all might be typing on a different keyboard and using Word-Perfect. As such, cultures, communities, even species, can get stuck optimizing to local conditions, rather than reach the global optimum. Indeed, one of Suarez's challenges to students was to ask them the degree to which they believe that they are actually free agents, or not completely in control of their own respective destinies. What local economies and cultures in their own backgrounds and families have led to them developing themselves as they exist at this moment? He also had students point to ways in which the choices they make and the people they become actually serve the interests of industries and economic orders. The agencies of those organizations, he pointed out, are distributed among us. Demerath chimed into say that in the same way we affirm and amend larger cultural orders based on gender, race, and sexuality as we both perform roles, and innovate within and across those roles. Students were reminded that the Hegelian dialectic used by Marx contains an element of feedback, as does the more recent theory of structuration by Giddens where innovated micro behavior can aggregate to change in macro phenomena which in turn constrains the micro level. The arrow of time first raised by Desjardins early in the course can be seen here at the level of society and cultural evolution.

As is often the case for undergraduate students in a sociology course, after taking in a thorough description of some social problem, there is a rising frustration that begs a question: what do we do about it? If it is true that we are often trapped in local pond of stagnation, and we are unknowingly sub-optimizing our environment, how can we persuade others that we can do better, that there are problems they don't see? Here is where a course like this calls for a professor comfortable with the idea that arts and literature are rhetorics of expression, communication, and persuasion. Enter Jeanne Hamming, who gave us poetry, art, science fiction, and nonfiction narratives as samples for realities that can be communicated but that are problematic, often caused by "suboptimization," to use Meadows' term.

Or should that be, "technological lock-in"? The overlaps in terminology at this point in the course begged the question, particularly from students eying the final exam: couldn't some streamlining take place? What is it? Suboptimization, technological lock-in, or if you want, over-exploitation, or is it something else? The best person to respond to that request for tidy order might well be any professor in arts and literature. The need for some chaos or disorder is not only what feeds creativity, but to over-categorize is to lose sight of the inimitable condition of that space in time, the phase state of that piece, that artist, that concept. Each of those concepts mean something a bit different. Suboptimization refers more to the environment, or network, while over-exploitation seems to chastise the blithely ignorant dominant agent, and lock-in disdains conservatism.

While Demerath wanted to clean up the scientific work bench and create a list of the most important concepts, Hamming did not. She was inclined to nix the antibiotic soap in an acknowledgement that extraneous differences within our communities are often crucial mediators between them and the outside world. "Dog owners get colds less often because they tend to have much higher microbial diversity, and that increases the likelihood that they'll be able to fight the occasional settlement of bad microbes. You want a diverse coalition." That's also the essence of Granovetter's "The Strength of Weak Ties" [32], one of the most cited articles in sociology, but mediators are critical for the growth of all complex systems. It has since become a line of argument Demerath has used to persuade sociology students that diversity is healthy for organizations and communities. Here, students accepted it as a reason to allow some latitude in the concept list. One of the thrills of the course was the awareness that we are all in on the ground floor of an emerging post-disciplinary field of study, so it's understandable that some of the chemical agents on the bench are labelled different ways, and not according to their attributes, again, but to their relationships with others agents.

Indeed, part of the reason complexity studies is itself emerging is because of the way it is able to "tolerate diversity" across the disciplines, being pragmatic about points of agreement, and overlooking differences in backgrounds. Hence the room for further distillation of concepts in our list that follows this section, but with the awareness that each discipline has its own values and truths that an overly condensed list could ignore or distort.

That pragmatism came to the fore again when philosopher Scott Davis rejoined us for the end of the course. A goal at that point was to consider ethics as emergent cultural norms that facilitate the growth of any collectivity. However, Davis assigned Harman's "Moral Relativism Defended" (2013) perhaps to challenge the class. Harman would not agree with Chaisson [11, 12], it would seem, that the universe is steadily increasing order and the accessibility of free energy. Nor would he agree with Kauffman, that we are seeing an evolution away from entropy. At least, if he does, he does not appear to see immoral behavior as that which acts against that tendency, or moral behavior as that which acts for it. That would be harmony, order, life's proliferation, complexity's continuation, as unpredictable and up-and-down as that might be.

But many in the class answered the call to resist Harman's source-less moral relativism, and Demerath fed them a heartening means of building towards a positive end to the course. This was Steven Pinker's *The Better Angels of Our Nature*, (2011) which shows a significant decline in violence over the course of human history. Most of us would be hard pressed to tolerate the violence that was routinely accepted in earlier epochs, and war is increasingly less common wherever you live today, as opposed to in past centuries and millennia. And an easy fit with this pattern is Durkheim's classic argument that people worship as sacred the support and security that they find in their community. Our sense of God, then, is a phenomenon that emerges from our sense of the power of our sociality to regulate and coordinate us; to order us and move us away from disorder and entropy. Where would we be without it? Hiding in woods behind a rock, more or less. No wonder we find conformity so sweet and good.

Interestingly, at the end of Pinker's Ted Talk describing the book's argument in 2007, he is inconclusive about why violence has declined. That should change with the rise of complexity studies, and the consensus that seems likely to emerge that the evolution away from entropy is real. Because order is on the rise, then, disruptive, chaotic experiences like war and violence are in decline. The emergence of higher order forms of organization such as government coalitions and the United Nations, as well as great networks of economic power in corporations, become increasingly committed to peace, with more and more at stake across parts of the world that could come into conflict. This consensus that peace is a dividend of complexity should not only be forming among those of us researching and teaching it, but also among the students who learn it as part of their basic undergraduate education.

8 Key Concepts

The course taught complexity concepts as they appear in each of their disciplinary discussions, as described in the previous section. The intent, however, was to always have students recognize, in implicit and explicit ways, how complexity concepts transcend each discussion and the discipline it inhabits. Every single concept has an application across more than one discipline, proving our point that complexity is interdisciplinary. The following is a short listing of key complexity concepts as well as brief description of how they were used in the course and its connection to our references and class readings. The order in which the concepts are presented roughly mimics the order in which the concepts were presented in our course.

Course concepts	Reasons for their significance
Systems theory	Historical precursor to complexity thinking, sharing the idea of different parts being linked to each other
Cellular automata	Simplest way of modeling agents reacting to local information while they interact in a lattice
Interaction and percolation	The interaction of independent parts gives rise emergent aggregate properties and patterns
Agent-based models	Showing how a few simple rules can result in the evolution of increasingly complex orders
Artificial life and multiagent systems	Interacting agents can be made increasingly realistic, be allowed to influence each other in unpredicted ways and give rise to circular causalities; because of this, these fields are often included in the area of Complex Adaptive Systems
Bottom-up approaches	Observed aggregate behavior is modeled as the emergent result of locally interacting parts (agents)
Self-organization of stars and cities	Noted by Christian as isomorphic

(continued)

(continued)

Course concepts	Reasons for their significance
Complexity as free energy	Charted by Chaisson, to increase over time through stars, life, brains and societies
Evolution away from entropy over time	Noted by Kauffman as a characteristic in the universe. Represents a core area of study in complexity
Homophily as attraction to shared information	In answer to Christian's call for a theory of social gravity to complement physical laws of attraction driving the increase of order
Energy and information	Information depends on energy; like energy, it also disperses; and agents are attracted to one another based on shared information
Non-linearity and emergence	Relationships change due to new information, leading to new patterns of equilibrium
Irreducibility	The algorithm is the shortest solution, meaning that a simulation is in itself an irreducible model
Phase states and phase transitions	Organization of information and energy define phase states; compression of information marks phase transitions
2nd law of thermodynamics	Seemingly at work in psychological and sociological phenomena as much as physical
Nested and co-evolving systems	Differentiation can occur externally or internally
Points of criticality at the edge of chaos	The disorder due to the 2nd law provides new information and gradients of difference to exploit
Evolution and punctuated equilibria	Where change can result from competition, random chance, death, memory loss, etc
Novelty in information and gradients of energy	Where novelty in a message between agents that share information is information, just as a difference in a slope
Cognition and aesthetic sensation	Thinking is satisfying for the subjective order it creates
Consciousness	By modeling the environment in our imagination we can focus on what is meaningful, allowing us to condense and process information more efficiently
Hierarchical structures and levels of reality	Simon stresses the fact that many aspects of nature and society are organized into hierarchies and multidimensional reality in a world with emergence is multifaceted
Power laws and scale-free networks	Hierarchical organization gives rise to recurring patterns in which aspects of reality are scale free
Positive and negative feedback loops	Interaction among agents gives rise to emergent patterns of behavior
"Fat tails" and the edge of chaos	Interaction among agents usually delivers more than a fair amount of extreme events because of feedback loops
Suboptimization	Self-interested action, where serving a dominant agent's interests limit or hurt the system as a whole
Technological lock-in	Accidents matter. The system may find its way to a local optimum (instead of the global one) and remain indefinitely there

(continued)

(continued)

Course concepts	Reasons for their significance
Exploitation and exploration	Complex optimization problems, from distributing valves in a steam engine to the best way to organize a society, where there are multiple local optima and a hidden global optimum can only be solved by trading off exploration and exploitation
Circular causality	Social science has struggled with ways of expressing and explaining the way in which individuals and aggregates interact; complexity thinking offers interesting ways of explaining these looped interactions
Rationality versus bounded rationality and low-cognition agents	Agents in traditional linear paradigms can be all-knowing and perfectly selfish decision makers, models with low-cognition agents assume away rationality and instead focus on the topology of interactions
Mediating agents	Gaining shared information through diverse ties mediates contact with potentially harmful agents
Agency	Agency in multi-leveled systems is bounded, both from above and from below

9 Conclusions

The purpose of this essay has been to convey the benefits of teaching complexity to undergraduate students as a way to help them obtain a broader view of their education, and how the different elements of what they are learning fit with each other. We are glad to have the opportunity to provide our students with a breadth of knowledge in many disciplinary fields. We may even brag about how our liberal arts graduates are so "well-rounded," have strong communication skills and capacity for critical thinking. We argue that what we do not focus enough attention is on teaching students how to synthesize the knowledge acquired into a coherent whole. Perhaps even more dramatically, students who are brought up in a disciplinary silo can graduate without understanding how their brick fits into the wall of science. More pragmatically, such a graduate may not possess the instructions describing how their knowledge, methodologies and techniques can be best accommodated in an interdisciplinary study, team or work environment.

The problems and tasks that our graduates will encounter throughout their careers will require them to see phenomena from a variety of perspectives, consider solutions to complex problems involving many different parts and constituents, participate in teams with different backgrounds, handle large amounts of quantitative as well as qualitative data, and integrate information from different sources in order to make wise decisions. We propose that complexity thinking can provide a theoretical backdrop in which to ground all of these aspects of decision-making in the modern world.

In this proposed interdisciplinary methodology, none of the participants will be completely comfortable understanding all aspects of the integrated approach, thus creating collaborators with a humbler approach to science and a nimbler understanding of the world. It also does not hurt that critical thinking skills are enhanced in the process, since each independent source of information will need to be carefully kneaded into the organized systemic approach. Finally, instead of being carpenters looking for nails to hammer, a complexity approach forces each disciplinary teammate to limit the way in which she advances the applicability of her disciplinary methodology, forcing her to evaluate the epistemology and applicability of her available tools.

The idea that interdisciplinarity can help students is not new. For example, Burchell [10] states that "*...Presumably exploration and investigation of knowledge that transcends disciplinary domains can enlarge the individual's capacity to build his own synthesis of knowledge and simultaneously can enable him to grapple with the unknown, thereby enhancing his power.*" What is novel about our approach is the use of complexity concepts as a door to such transdisciplinary approach to science and problem solving. In the process of working with diverse workmates, for example, a decision maker will have to deal with their inherent diversity, both in terms of their socio-cultural backgrounds as well as in terms of their disciplinary upbringings. A well-versed manager or collaborator will thus have to translate the information he receives from contrasting sources in order to successfully integrate a coherent plan of action.

The concept of emergence, so integral to complexity thinking, explains why disciplines exist and why they have developed their own languages to describe reality. While it is fairly undisputable that we will always have disciplines, the future is now pointing to a more holistic approach to science, potentially based around a new, emerging paradigm of complexity. Emergence both gives rise to disciplines and provides a bridge to reintegrate the seemingly disjointed bodies of knowledge. True interdisciplinarity merges diverse approaches to science, giving birth to a unique analysis that is in itself an emergent phenomenon that is more than the sum of its parts.

Students in the 21st century may often find themselves owning siloed disciplinary understandings of the world around them, as well as methodological tools that are at first glance incompatible with each other. Complexity opens the door to transdisciplinarity, allowing us to undo a potentially arbitrary division of isolated disciplinary knowledge into bespoke approaches to understanding a particular phenomenon, creating new and powerful tools and methodologies for solving real-world problems. The undergraduate complexity course proposed in this essay calls for creativity, granting the student permission to think about an issue from scratch, developing her own particular toolkit to appropriately analyze any aspect of reality, and tackle the increasingly difficult and interconnected problems of our world.

References

1. Abbott, R.: Emergence explained: abstractions: getting epiphenomena to do real work. Complexity **12**(1), 13–26 (2006)
2. Aminian, F., Suarez, E.D., Aminian, M., Walz, D.T.: The benefits of modeling economic data with neural networks. J. Comp. Econ. **28**(1) (2006)
3. Astin, A.: How the Liberal arts college affects students. Daedalus **128**(1), 77–100 (1999)
4. Bak, P., Chen, K., Creutz, M.: Self-organized criticality in the game of life. Nature **342**(6251), 780–782 (1989)
5. Bammer, G., Bronitt, S., Brown, L., Bursztyn, M., Maury, M., Cram, L., Withers, G.: The relationship of integrative applied research and I2S to multidisciplinarity and transdisciplinarity. In: Disciplining Interdisciplinarity: Integration and Implementation Sciences for Researching Complex Real-World Problems, pp. 213–220. ANU Press (2013)
6. Bar-Yam, Y.: A mathematical theory of strong emergence using multiscale variety. Complexity **9**(6) (2004)
7. Binder, F.: Interdisciplinary courses. Improv. Coll. Univ. Teach. **7**(2), 45–47 (1959)
8. Brown, D.G., Page, S., Riolo, R., Zellner, M., Rand, W.: Path dependence and the validation of agent-based spatial models of land use. Int. J. Geogr. Inf. Sci. **19**(2), 153–174 (2005)
9. Brown, D.G., Robinson, D.T., An, L., Nassauer, J.I., Zellner, M., Rand, W., Wang, Z.: Exurbia from the bottom-up: confronting empirical challenges to characterizing a complex system. Geoforum **39**(2), 805–818 (2008)
10. Burchell, H.: On interdisciplinary education. High Sch. J. **55**(2), 78–85 (1971)
11. Chaisson, E.: Cosmic Evolution: The Rise of Complexity in Nature. Harvard University Press, Cambridge (2001)
12. Chaisson, E.: Energy rate density II: probing further a new complexity Metric. Complexity **17**, 44–63 (2011)
13. Christian, D.: Maps of Time: An Introduction to Big History. University of California Press, Berkeley (2004)
14. Collell, G., Fauquet, J.: Brain activity and cognition: a connection from thermodynamics and information theory. Front. Psychol. **6**, 818 (2015)
15. Cortese, A.: The critical role of higher education in creating a sustainable future. High. Educ. 15–22 (2003)
16. Damuth, J., Heisler, I.L.: Alternative formulations of multilevel selection. Biol. Philos. **3**(4), 407–430 (1988)
17. Damuth, J.D.: Population ecology: common rules for animals and plants. Nature **395**(6698), 115 (1998)
18. Danks, D.: Unifying the Mind: Cognitive Representations as Graphical Models. MIT Press, Cambridge (2014)
19. De Greef, L., Post, G., Vink, C., Wenting, L.: Designing interdisciplinary education: a practical handbook for university teachers. Chapter 3: Unravelling Interdisciplinary Understanding, pp. 28–42. Amsterdam University Press, Amsterdam (2017)
20. De Landa, M.: A thousand years of nonlinear history. Swerve Editions (2000)
21. De Landa, M.: A new philosophy of society: assemblage theory and social complexity. Continuum (2006)
22. Demerath, L.: Epistemological culture theory: a micro account of the origin and maintenance of culture. Sociol. Theory **20**(2), 208–226 (2002)
23. Demerath, L.: Explaining culture: the social pursuit of subjective order. Lexington, Lanham NJ (2012)
24. Demerath, L.: Knowledge-based affect: cognitive origins of "Good" and "Bad". Soc. Psychol. Q. 136–147 (1993)
25. Eckley, R.: Liberal arts colleges: can they compete? Brook. Rev. **5**(4), 31–37 (1987)
26. Edmonds, B., Moss, S.: From KISS to KIDS – an "anti-simplistic" modelling approach. In: P. Davidsson, B. Logan, B. Takadama, K. (eds.) Multi-agent and Multi-Agent-Based Simulation, Lecture Notes In Artificial Intelligence, pp. 130–144. Springer, Berlin (2005)

27. Epstein, J.M.: Why model?. *J. Artif. Soc. Soc. Simul.* **11**(4) 12 (2008)
28. Geels, F.: Co-evolution of technology and society: the transition in water supply and personal hygiene in the Netherlands (1850–1930)—a case study in multi-level perspective. Technol. Soc. **27**(3), 363–397 (2005)
29. Gersick, C.J.: Revolutionary change theories: a multilevel exploration of the punctuated equilibrium paradigm. Acad. Manag. Rev. **16**(1), 10–36 (1991)
30. Goldt, S., Seifert, U.: Stochastic thermodynamics of learning. Phys. Rev. Lett. **118**(1), 010601 (2017)
31. Goodwin, B.: Beyond the darwinian paradigm: understanding biological forms. In: Ruse, M., Travis, J. (eds.) Evolution: The First Four Billion Years, pp. 299–312. Harvard University Press, Cambridge (2009)
32. Granovetter, M.: The strength of weak ties. Am. J. Sociol. **78**(6), 1360–1380 (1973)
33. Harman, G.: Moral relativism defended. Ethical Theor. Anth. 35–43 (2013)
34. Hatt, K.: Considering complexity: toward a strategy for non-linear analysis. Can. J. Sociol. **34**(2), 313–347 (2009)
35. Hill, W.: Interdisciplinary perspectives and the liberal arts. In: Chopp, R., Frost, S., Weiss, D.H. (eds.) Remaking College: Innovation and the Liberal Arts, pp. 85–95. Johns Hopkins University Press, Baltimore (2013)
36. Holland, J.H.: Complex adaptive systems. Daedalus 17–30 (1992)
37. Holland, J.H.: Complexity: a very short introduction. OUP Oxford (2014)
38. Jacobs, J., Frickel, S.: Interdisciplinarity: a critical assessment. Ann. Rev. Sociol. **35**, 43–65 (2009)
39. Johnson, N.F.: Two's company, three is complexity: a simple guide to the science of all sciences. Oneworld Pubns Ltd (2007)
40. Johnson, S.: Emergence: The Connected Lives of Ants, Brains, Cities, p. 19. Scribner, New York (2001). ISBN 3411040742
41. Kahneman, D.: Thinking, Fast and Slow. Macmillan (2011)
42. Katz, P.M.: Interdisciplinary Undergraduate Education: CIC Project on the Future of Independent Higher Education. Innovations in Teaching and Learning. Research Brief 2 (2015)
43. Kauffman, S.: At home in the universe: the search for the laws of self-organization and complexity. Oxford university press, Oxford (1996)
44. Kaufman, B.: In: Emmett, R.B. (ed.) The Elgar Companion to the Chicago School of Economics, p. 133 (2010)
45. Koch, C.: Consciousness: Confessions of a Romantic Reductionist. MIT Press, Cambridge (Massachusetts) (2012)
46. Latour, B.: We Have Never Been Modern. Harvard university press (2012)
47. Lenartowicz, M., Weinbaum, D., Braathen, P.: Social systems: complex adaptive loci of cognition. Emerg.: Complex. Organ. **18**(2) (2016)
48. Lewis, J.D., Weigart, A.J.: The structures and meanings of social time. Soc. Forces **60**(2), 432–462 (1981)
49. Lind, M.: Why the liberal arts still matter. Wilson Q. (1976), **30**(4), 52–58 (2006)
50. Luhmann, N.: Social Systems. Stanford University Press, Stanford (1996)
51. McCool, S., Freimund, W., Breen, C., Gorricho, J., Kohl, J., Biggs, H.: Benefiting from complexity thinking. In: Worboys, G., Lockwood, M., Kothari, A., Feary, S., Pulsford, I. (Eds.), Protected Area Governance and Management pp. 291–326. ANU Press (2015)
52. Miller, J.H., Page, S.E.: The standing ovation problem. Complexity **9**(5), 8–16 (2004)
53. Moore, T., Roberts, R.: Statistics at liberal arts colleges. Am. Stat. **43**(2), 80–85 (1989)
54. Nicolescu, B.: Methodology of transdisciplinarity. World Futur. J. New Parad. Res. **70**(3–4) Transdisciplinarity (2014)
55. Nicolis, G., Prigogine, I.: Exploring complexity: an introduction. Freeman, New York (1989)
56. Nowacek, R.: A discourse-based theory of interdisciplinary connections. J. Gen. Educ. **54**(3), 171–195 (2005)
57. Nowotny, H.: The increase of complexity and its reduction emergent interfaces between the natural sciences, humanities and social sciences. Theory, Cult. Soc. **22**(5), 15–31 (2005)

58. Page, S.: The Difference: How the Power of Diversity Creates Better Groups, Firms, Schools, and Societies (New Edition). Princeton University Press, Princeton, Oxford (2007)
59. Pinker, S.: The Better Angels of Our Nature: The Decline of Violence in History and its Causes. Penguin uk (2011)
60. Reber, R., Schwarz, N., Winkielman, P.: Processing fluency and aesthetic pleasure: is beauty in the perceiver's processing experience? Pers. Soc. Psychol. Rev. **8**(4), 364–382 (2004)
61. Resnicow, K., Page, S.E.: Embracing chaos and complexity: a quantum change for public health. Am. J. Public Health **98**(8), 1382–1389 (2008)
62. Reybold, L.Earle, Halx, Mark D.: Coming to terms with the meaning of interdisciplinarity: faculty rewards and the authority of the discipline. J. Gen. Educ. **61**(4), 323–351 (2012)
63. Rutting, L., Post, G., De Roo, M., Blad, S., De Greef, L.: An Introduction to Interdisciplinary Research: Theory and Practice, Menken, S., Keestra M. (Eds.) Chapter 6: Interdisciplinary integration, pp. 41–49, Amsterdam University Press, Amsterdam (2016)
64. Sawyer, R.K.: Social emergence: Societies as complex systems. Cambridge University Press, Cambridge (2005)
65. Seifert, T., Goodman, K., Lindsay, N., Jorgensen, J., Wolniak, G., Pascarella, E., Blaich, C.: The effects of liberal arts experiences on liberal arts outcomes. Res. High. Educ. **49**(2), 107–125 (2008)
66. Simon, Herbert: The Sciences of the Artificial, 3rd edn. MIT Press Cambridge, MA (1996). ISBN 0-262-69191-4
67. Simondon, G.: The Genesis of the Individual. In: Zone: Incorporations 6, Crary, J., Kwinter, S. (eds.) Zone, New York (1992) ISBN: 0942299299
68. Suarez, E.D., Castañón-Puga, M.: Distributed agency. Int. J. Agent Technol. Syst. **5**(1), 32–52 (2013)
69. Swora, T., Morrison, J.: Interdisciplinarity and higher education. J. Gen. Educ. **26**(1), 45–52 (1974)
70. Tarde, G.: The Laws of imitation. Holt, H. (1903)
71. Tononi, G.: Consciousness as integrated information: a provisional manifesto. Biol. Bull. **215**(3), 216–242 (2008)
72. Waldrop, M.M.: Complexity: The Emerging Science at the Edge of Order and Chaos. Touchstone (1992)
73. Waldrop, M.M.: Complexity: The Emerging Science at the Edge of Order and Chaos, Simon and Schuster (1993)
74. Wegner, P.: Interaction as a basis for empirical computer science. [CSUR]. ACM Comput. Surv. **27**(1), 45–48 (1995)
75. Wilson, Edward O.: Consilience: the unity of knowledge. Vintage Books/Random House Inc, New York (1999)
76. Wilson, D.S., Wilson, E.O.: Rethinking the theoretical foundation of sociobiology. Q. Rev. Biol. **82**(4), 327–348 (2007)
77. Wilson, D.S., Van Vugt, M., O'Gorman, R.: Multilevel selection theory and major evolutionary transitions: implications for psychological science. Curr. Dir. Psychol. Sci. **17**(1), 6–9 (2008)

GPSR Compliance

The European Union's (EU) General Product Safety Regulation (GPSR) is a set of rules that requires consumer products to be safe and our obligations to ensure this.

If you have any concerns about our products, you can contact us on ProductSafety@springernature.com

In case Publisher is established outside the EU, the EU authorized representative is:

Springer Nature Customer Service Center GmbH
Europaplatz 3
69115 Heidelberg, Germany

Batch number: 10777861

Printed by Printforce, the Netherlands